Jenny Head

£7·95

Applied Statistics

PRINCIPLES AND EXAMPLES

D1332671

Applied Statistics
PRINCIPLES AND EXAMPLES

D.R. COX
E.J. SNELL
Department of Mathematics,
Imperial College, University of London

LONDON   NEW YORK

CHAPMAN AND HALL

First published 1981 by
Chapman and Hall Ltd
11 New Fetter Lane, London EC4P 4EE
Reprinted 1982

Published in the USA by
Chapman and Hall
in association with Methuen, Inc.
733 Third Avenue, New York NY 10017

© 1981 D.R. Cox and E.J. Snell

Printed in Great Britain
at the University Press, Cambridge

ISBN 0 412 16560 0 (cased)
ISBN 0 412 16570 8 (paperback)

This title is available in both hardbound and paperback editions. The paperback edition is sold subject
to the condition that it shall not, by way of trade
or otherwise, be lent, re-sold, hired out, or
otherwise circulated without the publisher's prior
consent in any form of binding or cover other than
that in which it is published and without a similar
condition including this condition being imposed on
the subsequent purchaser.

All rights reserved. No part of this book may be
reprinted, or reproduced or utilized in any form or
by any electronic, mechanical or other means, now
known or hereafter invented, including photocopying
and recording, or in any information storage and
retrieval system, without permission in writing from
the Publisher.

British Library Cataloguing in Publication Data

Cox, D.R.
  Applied statistics.

  1. Mathematical statistics
  I. Title   II. Snell, E.J.

  519.5   QA276
  ISBN 0-412-16560-0
  ISBN 0-412-16570-8 Pbk

Contents

Preface

There are many books which set out the more commonly used statistical methods in a form suitable for applications. There are also widely available computer packages for implementing these techniques in a relatively painless way. We have in the present book concentrated not so much on the techniques themselves but rather on the general issues involved in their fruitful application.

The book is in two parts, the first dealing with general ideas and principles and the second with a range of examples, all, however, involving fairly small sets of data and fairly standard techniques. Readers who have experience of the application of statistical methods may want to concentrate on the first part, using the second part, and better still their own experience, to illuminate and criticize the general ideas. If the book is used by students with little or no experience of applications, a selection of examples from the second part of the book should be studied first, any general principles being introduced at a later stage when at least some background for their understanding is available.

After some hesitation we have decided to say virtually nothing about detailed computation. This is partly because the procedures readily available will be different in different institutions. Those having access to GLIM will find that most of the examples can be very conveniently handled; however the parameterization in GLIM, while appropriate for the great generality achieved, is not always suitable for interpretation and presentation of conclusions. Most, although not all, of the examples are in fact small enough to be analysed on a good pocket calculator. Students will find it instructive themselves to carry out the detailed analysis.

We do not put forward our analyses of the examples as definitive. If the examples are used in teaching statistical methods, students should be encouraged to try out their own ideas and to compare thoughtfully the conclusions from alternative analyses. Further sets of data are included for use by students.

Many of the examples depend in some way on application of the method of least squares or analysis of variance or maximum likelihood. Some familiarity with these is assumed, references being given for specific points.

The examples all illustrate real applications of statistical methods to some branch of science or technology, although in a few cases fictitious data have been supplied. The main general limitation on the examples is, as noted above, that inevitably they all involve quite small amounts of data, and important aspects of statistical analysis specific to large amounts of data are therefore not well covered. There is the further point that in practice over-elaboration of analysis is to be avoided. With very small sets of data, simple graphs and summary statistics may tell all, yet we have regarded it as legitimate for illustration in some cases to apply rather more elaborate analyses than in practice would be justified.

We are grateful to Dr C. Chatfield, University of Bath, for constructive comments on a preliminary version of the book.

D.R. Cox
E.J. Snell

London, September 1980

Part I   Principles

Chapter 1

Nature and objectives of statistical analysis

1.1 Introduction

Statistical analysis deals with those aspects of the analysis of data that are not highly specific to particular fields of study. That is, the object is to provide concepts and methods that will, with suitable modification, be applicable in many different fields of application; indeed one of the attractions of the subject is precisely this breadth of potential application.

This book is divided into two parts. In the first we try to outline, without going into much specific detail, some of the general ideas involved in applying statistical methods. In the second part, we discuss some special problems, aiming to illustrate both the general principles discussed earlier and also particular techniques. References to these problems are given in Part I where appropriate. While all the examples are real, discussion of them is inhibited by two fairly obvious constraints. Firstly, it is difficult in a book to convey the interplay between subject-matter considerations and statistical analysis that is essential for fruitful work. Secondly, for obvious reasons, the sets of data analysed are quite small. In addition to the extra computation involved in the analysis of large sets of data, there are further difficulties connected, for example, with its being hard in large sets of data to detect initially unanticipated complications. To the extent that many modern applications involve large sets of data, this book thus paints an oversimplified picture of applied statistical work.

We deal very largely with methods for the careful analysis and interpretation of bodies of scientific and technological data. Many of the ideas are in fact very relevant also to procedures for decision making, as in industrial acceptance sampling and automatic process control, but there are special issues in such applications, arising partly from the relatively mechanical nature of the final procedures. In many applications, however, careful consideration of objectives will indicate a specific feature of central interest.

Special considerations enter also into the standardized analysis of routine test data, for example in a medical or industrial context. The need here may be for clearly specified procedures that can be applied, again in a quite mechanical fashion, giving sensible answers in a wide range of circumstances, and allowing possibly for individual 'management by exception' in extremely

3

peculiar situations; quite commonly, simple statistical procedures are built in to measuring equipment. Ideally, largely automatic rejection of 'outliers' and routine quality control of measurement techniques are incorporated. In the present book, however, we are primarily concerned with the individual analysis of unique sets of data.

1.2 Data quality

We shall not in this book deal other than incidentally with the planning of data collection, e.g. the design of experiments, although it is clear in a general way that careful attention to design can simplify analysis and strengthen interpretation.

We begin the discussion here, however, by supposing that data become available for analysis. The first concerns are then with the quality of the data and with what can be broadly called its structure. In this section we discuss briefly data quality.

Checks of data quality typically include:

(i) visual or automatic inspection of the data for values that are logically inconsistent or in conflict with prior information about the ranges likely to arise for the various variables. For instances of possibly extreme observations, see Examples E and S. Inspection of the minimum and maximum of each variable is a minimal check;

(ii) examination of frequency distributions of the main variables to look for small groups of discrepant observations;

(iii) examination of scatter plots of pairs of variables likely to be highly related, this detecting discrepant observations more sensitively than (ii);

(iv) a check of the methods of data collection to discover the sources, if any, of biases in measurement (e.g. differences between observers) which it may be necessary to allow for in analysis, and to assess the approximate measurement and recording errors for the main variables;

(v) a search for missing observations, including observations that have been omitted because of their highly suspicious character. Often missing observations are denoted in some conventional way, such as 0 or 99, and it will be important not to enter these as real values in any analysis.

Concern that data quality should be high without extensive effort being spent on achieving unrealistically high precision is of great importance. In particular, recording of data to a large number of digits can be wasteful; on the other hand, excessive rounding sacrifices information. The extent to which poor data quality can be set right by more elaborate analysis is very limited, particularly when appreciable systematic errors are likely to be present and cannot be investigated and removed. By and large such poor-quality data will not merit very detailed analysis.

In nearly all the problems we shall be dealing with there will be appreciable uncontrolled variation. Our attitude to this will vary a little depending on whether the variability is natural variation that is an intrinsic part of the system under study or on whether it represents error of measurement or lack of control that could in principle be eliminated. In both cases we consider the frequency distribution of the variation. In the former case we may well be interested in detail in the form of the distribution; this applies, for instance, to the survival times of Example U. In the second case, interest will ultimately be in the basic constants under study, i.e. in what would have been observed had 'error' been eliminated. Example O, concerned with the determination of a physical constant, is an extreme example.

When the amount of data is large, the recording and storage of data in a form that makes analysis relatively easy become very important. Thus recording directly on punched cards or magnetic tape may be called for. In other cases, especially in the physical sciences, it is common for digital, or less often analogue, computation of summarizing quantities to be part of the measuring equipment. While digital computers have greatly extended the possibilities for the analysis of data, it is important to realize that, at least in some fields, they have produced an even greater increase in the amount of data that can be obtained and recorded, the so-called 'information explosion'. The relative ease with which data can be obtained and analysed varies greatly between different fields and has important implications for the depth of analysis that is sensible.

1.3 Data structure and quantity

Most, although not all, data have the following broad form. There are a number of individuals (people, plots, experimental animals, etc.) and on each individual a number of types of observation are recorded. Individuals are thought to be in some sense independent of one another.

The following questions then arise:

(i) what is to be regarded as an individual?

(ii) are the individuals grouped or associated in ways that must be taken account of in analysis?

(ii) what are the variables measured on each individual?

(iv) are any observations missing, and if so, what can be done to replace or estimate those values?

Data structure is thus a question partly of the number and nature of the variables measured on each individual, and partly of the classification and groupings of individuals. The quantity of data is best thought of as having two aspects, the number of individuals and the number of variables per individual. As already noted, there is a qualitative difference between situations, like those in Part II, in which the total amount of data is such that it can all be carefully inspected during analysis and, at the other extreme, applications in which the amount of data is so vast that at most a small proportion of it can be analysed.

1.4 Phases of analysis

It is convenient to distinguish four broad phases to statistical analysis. These are:

(i) initial data manipulation, i.e. the assembling of the data in a form suitable for detailed analysis and the carrying out of checks on quality of the kind outlined in Section 1.2;

(ii) preliminary analysis, in which the intention is to clarify the general form of the data and to suggest the direction which a more elaborate analysis may take. Often this is best done by simple graphs and tables;

(iii) definitive analysis, which is intended to provide the basis for the conclusions;

(iv) presentation of conclusions in an accurate, concise and lucid form. This leads usually to a subject-matter interpretation of the conclusions, which, while obviously crucial, we regard as outside the present discussion.

While this division is useful, it should not be taken rigidly. Thus an analysis originally intended as preliminary may give such clear results that it can be regarded as definitive. Equally, an analysis intended as definitive may reveal unexpected discrepancies that demand a reconsideration of the whole basis of the analysis. In fields in which there is substantial experience of previous similar investigations one may hope largely to bypass preliminary analysis.

Clear presentation of conclusions is of great importance; the style to be adopted of course depends to some extent on the potential audience. It is virtually always important that the broad strategy of the analysis is explained in a form that seems reasonable to a critical nontechnical reader.

These points imply that conceptually simple methods are to be preferred to conceptually complicated ones. In particular, direct links between the final conclusions and the data are a good thing. This is partly because presentation of conclusions is eased and partly because sensitivity of the conclusions to assumptions is more readily assessed. Thus in Example L we have chosen not to transform some proportions on the grounds that the conclusions are more easily appreciated directly.

Effort spent in trying to present in a simple way the conclusions of complex analyses is almost always worth while. On the other hand, simplicity is to some extent a subjective feature strongly correlated with familiarity. Understanding of a particular technique comes partly from fruitful application of it. Thus, time permitting, it can be valuable to try out as yet unfamiliar techniques on reasonably well understood sets of data, as essentially a training exercise rather than as part of the immediate scientific analysis.

1.5 Styles of analysis

Methods of analysis can be grouped in several broad ways. Firstly, we may distinguish between descriptive methods and probabilistically based methods. In the latter, but not the former, the notion of a probability model for the data and often of the probabilistic properties of estimates and their uncertainty occur explicitly. In descriptive statistics the probabilistic aspects are either absent or at least receive little emphasis. Many of the most valuable methods, however, are sensible from both points of view. If, as is often the case, explicit consideration of the uncertainty in the conclusions is required, probabilistic methods are likely to be needed.

A second broad distinction is between graphical and numerical techniques. Definitive probabilistic analysis is virtually always numerical rather than graphical, but even with these methods graphical methods can be valuable in presenting conclusions. In other types of analysis, there will often be a role for both graphical and numerical techniques.

Graphical methods are of much value in presenting qualitative aspects of conclusions in an easily grasped way. Nevertheless, it will almost always be desirable to give the main summarizing quantities numerically, together with suitable measures of precision, such as estimated standard errors, or at least the information from which such measures of precision can be calculated. One reason for requiring numerical statements is that future workers wishing to use the results will often have convenient access only to the reported conclusions. Reconstruction of numerical information from published graphs nearly always introduces further errors. More broadly, especially in preparing papers for scientific journals, one should consider the needs of future workers in the field, who may wish to use the results in unexpected ways.

1.6 Computational and numerical analytical aspects

Two important aspects of statistical analysis about which rather little will be said in this book are the numerical analytic and computational. Numerical analytic considerations arise in ensuring that procedures are not sensitive to rounding errors in calculation and that where iterative procedures are necessary they converge, preferably speedily, to the required point. Example F illustrates the possibility that an apparently simple analysis may be subject to major rounding errors.

Computational considerations are broader and involve the organization of the raw data and the arrangement of the final conclusions, as well as the implementation of the main analysis. The wide availability of computers means that larger sets of data can be handled, that more ambitious methods of analysis can be used and that, once a particular kind of analysis is programmed, the calculations can be repeated many times in slightly different forms without much extra work. Program packages are available for most of the more standard statistical calculations, although these sometimes lack the flexibility that ideally should be allowed. For most of the small sets of data and relatively simple analyses contemplated in Part II of this book, an electronic pocket calculator is adequate.

For very large-scale investigations with large amounts of data, a substantial effort is likely to be necessary on data recording and storage. It is thus justifiable to devote, if necessary, some time to the development of special programs for analysis. In principle it is, of course, desirable that the planning of the recording and of the analysis of the data should be coordinated. For investigations with a small amount of data of relatively simple structure, computational considerations will not normally be critical. For what is probably the most common occurrence, data of moderate size with limited resources available for analysis, major development of special programs for analysis is not feasible and the availability of flexible and general program packages is of central importance. The temptation to use elaborate methods of analysis just because programs are available is, however, to be avoided, except as a training exercise.

Because computational considerations depend so much on the resources available and on rapidly developing technology, we shall in this book say little on the subject. Of its great importance there is, however, no doubt.

1.7 Response and explanatory variables

A rough classification of kinds of observation will be given in Section 2.1, but one distinction is sufficiently important to mention immediately. This is between response variables and explanatory variables. In any particular section of analysis, we typically regard one or more variables as responses and other variables as explanatory variables and consider the question: how does the response variable depend on the explanatory variable? The distinction is most clear cut when the explanatory variables represent treatments imposed by the investigator, and it is required to assess the effect of treatment on some subsequent response; for example, the response variable may be the yield from a chemical reaction and the explanatory variables may specify the temperature and pressure of the reaction, the concentrations of reactants, etc. More broadly, the response variables are in some general sense regarded as dependent on the explanatory variables; indeed, dependent variable is an alternative name for response variable. Sometimes the response variable is so chosen because the ultimate object is to predict it from the explanatory variables. For two social science problems where the correct isolation of response from explanatory variables is important and not straightforward, see Examples W and X.

The recognition of response and explanatory variables is one of the first steps in clarifying the approach to analysis. Often in practice there are several response variables, e.g. yields of several products from a chemical reaction, or a whole frequency distribution, as in Example C. Analysis is much simplified if the individual response variables can, at least for detailed analysis, either be dealt with entirely separately or a single combined response variable formed, as in Examples C and M. Multivariate analysis is that part of statistical analysis dealing with several response variables simultaneously and tends to give conceptually fairly complicated results. Its use is not illustrated in the present book. Fruitful use is mostly in fields where many response variables of a rather similar kind are recorded and reduction of dimensionality is essential.

In essence, then, response variables are the primary properties of interest (yield of product, success or failure of medical treatment, energies and lifetimes of particles in nuclear physics, etc.). Explanatory variables (dose level, treatment applied, etc.) hopefully explain systematic variations in the response variables.

Sometimes it is useful to distinguish intermediate response variables, which are broadly response variables which in some stages of analysis and interpretation may be treated as explanatory variables. For instance, in an agricultural field trial, the main response variable may be yield per m^2, whereas an intermediate response variable may be a number of plants per m^2. In a medical trial, say on the relief of pain, the main response variable may be a subjective score of relief achieved and an intermediate response variable may be change in blood pressure or some biochemical variable. In both examples, it would be possible to analyse the intermediate response variable as a response, assessing, for example, the effect of treatments on it. If, however, we use the intermediate response as an explanatory variable we are addressing the more subtle question as to the extent to which the effect of treatment on the main response variable is accounted for by the action of the intermediate variable.

It is important throughout the discussion of analysis that a variable, whether response or explanatory, may well have been produced by an initial process of combination of raw observations. For instance, in a sugar-beet experiment an important response variable may be effective sugar yield in kg/m^2, derived from measurements on yield of beet and on chemical analysis of sugar content. A more elaborate example concerns growth curves, say in an animal feeding experiment. Here possible response variables are: (i) live weight after some fixed time, measured directly; (ii) an estimate of asymptotic weight; (iii) some estimate of rate of approach to that asymptote; (iv) some measure of the efficiency of conversion of food into body weight. The last three would all involve nontrivial intermediate analysis of the original observations. Where necessary such new variables will be called derived variables. Examples C, E and M all illustrate simple instances of the formation of a derived variable, in the first and last by combination of a set of frequencies into a single summarizing quantity.

1.8   Types of investigation

The design of investigations will not be considered in detail in the present book; nevertheless, it is desirable from the point of view of interpretation to distinguish a number of types of investigation. While the same technique of analysis, e.g. regression analysis, may be used in numerically identical form for any one of the main types of study, the limitations on scientific interpretation are quite different in the different types. It is helpful to distinguish between the following:

(i) Experiments, in which the system under study is set up and controlled by the investigator. Typically, one of a number of alternative treatments is applied to each individual, or experimental unit, and responses measured. If the allocation of treatments to experimental units is organized by the investigator, and especially if an element of objective randomization is involved, it will be possible to conclude that any clear-cut difference in response between two treatments is a consequence of the treatments.

(ii) Pure observational studies, in which data have been collected on individuals in some system, the investigator having had no control over the collection of the data, other than perhaps some role in checking the quality of the data. While it may be possible to detect from such data clear effects, such as differences between different groups of individuals, interpretation of such differences will nearly always call for much caution. Explanatory variables that would provide the 'real' explanation of the differences may not have been measured, and may even be unknown to the investigator.

(iii) Sample surveys, in which a sample is drawn from a well-defined population by methods, usually involving randomization, under the investigator's control. Conclusions can be drawn with confidence about the descriptive properties of the population in question, but the interpretation of, for example, relationships between variables raises problems similar to (ii). Control of data quality may be stronger than in a pure observational study.

(iv) Controlled prospective studies, in which a group of individuals, chosen by the investigator, have various explanatory variables measured and are then followed through time, often to see whether some particular event of significance (e.g. death) occurs. To the extent that all important explanatory variables can be measured, and of course this is never totally possible, these studies have some of the virtues of an experiment.

(v) Controlled retrospective studies, such as the epidemiological investigation summarized in Example V, in which a characteristic response variable has been observed on individuals, and the history of those individuals is examined to isolate relevant explanatory variables.

Experiments are strongly interventionist, the investigator having in principle total control over the system under study, and lead to the clearest interpretation. While very often experiments can be arranged to lead to simple 'balanced' sets of data, this is not the crucial point. In principle virtually any method of statistical analysis might be relevant for any style of investigation; it is the interpretation that differs.

An outline example will clarify the distinction between an experiment and a pure observational study. Consider the comparison of two alternative medical treatments A and B. In an experiment each eligible patient is assigned a treatment by an objective randomization procedure, each patient having equal chance of receiving each treatment. Suppose that subsequent patient care is closely standardized, identically for the two treatments, and that a clearly defined response variable is measured for each patient, e.g. survival for at least one year versus death within one year. Suppose that, as judged by an appropriate statistical procedure, the two groups of response variables differ by more than can reasonably be ascribed to chance. We can then conclude that, provided the experiment has been correctly administered and reported, the difference between the groups is a consequence of the difference between A and B; for the two groups of patients differ only by the accidents of random assignment and in virtue of the difference between A and B.

Contrast this with a pure observational study in which, from hospital records, information is assembled on the same response variable for two groups of patients, one group having received treatment A and the other treatment B. Note first that the structure of the data might be identical for the experiment and for the observational study. Suppose that again there is a clear difference between the two groups. What can we conclude?

The statistical analysis shows that the difference is unlikely to be a pure chance one. There are, however, initially many possible explanations of the difference in addition to a possible treatment effect. Thus the groups may differ substantially in age distribution, sex, severity of initial symptoms, etc. Specific explanations of this kind can be examined by suitable statistical analysis, although there always remains the possibility that some unmeasured explanatory variable differs very substantially between the two groups. Further, we rarely know why each patient was assigned to his or her treatment group: the possibility that conscious or unconscious assessment of the patient's prognosis influenced treatment choice can rarely be excluded. Thus the interpretation of the difference is more hazardous in the observational study than in the experiment.

In the light of these distinctions it is quite often useful to divide explanatory variables into two types. First there are those which represent, or which could conceivably have represented, treatments. Then there are those that give intrinsic properties of the individuals; we call the latter type intrinsic variables.

For example, suppose that in a medical investigation the explanatory variables are dose of a drug, patient sex and patient initial body weight, response being some measure of success of treatment. Now analysis would usually be directed towards relating response to dose. The role of body weight might be to indicate an appropriate scale for dose, e.g. mg per kg body weight, or to show how the response–dose relation is modified by body weight. Similarly, it may be necessary to estimate different response–dose relations for men and for women. It would usually be meaningful to think of dose as causing response, because it is possible to contemplate an individual receiving a different dose from the one he or she in fact did receive. But it would be bad terminology, if no more, to say that patient sex 'causes' a difference in response, even were it to happen that the only systematic difference found in the data were a difference between men and women. This is because it is not usually meaningful to contemplate what response would have been observed on an individual had that individual been a woman rather than a man. The point is related to the physicists' well-known dictum that passage of time cannot be regarded as a cause of change.

1.9 Purposes of investigation

In the previous section we have classified investigations by their design, the primary distinction being between experimental and observational studies. A rather different division of investigations can be made on the basis of their broad purpose. In one sense, of course, it is trite to remark that the purpose of the investigation is to be borne in mind, particularly in determining the primary aspects of the model. Indeed, in some applications the objectives may be very specific and of such a kind that the quantitative techniques of decision analysis may be applicable.

Nevertheless, it is useful to draw a broad qualitative distinction between investigations, or parts of investigations, whose objective is in some sense to increase understanding and those with a much more specific 'practical' objective. The terms 'scientific' and 'technological' might be used. We shall prefer 'explanatory' and 'pragmatic', partly to avoid the misunderstanding that the subject matter is at issue. Thus an investigation in nuclear physics to calibrate a technique or to compare alternative experimental procedures might be severely pragmatic, whereas an experiment on alternative animal management techniques in agriculture might be explanatory, being set up to give understanding of the biological reasons for differences, and not just a determination of which is the economically better technique in the context studied.

The distinction has bearing on the kinds of conclusion to be sought and on the presentation of the conclusions. For example, a pragmatic application of multiple regression might aim to predict one or more response variables from suitable explanatory variables. Then if there were a number of alternative predicting equations giving about equally good results, the choice between them could be made on grounds of convenience or even essentially arbitrarily. If, however, the objective is the understanding of the relation between the response variable and the explanatory variables, interest will lie in which explanatory variables contribute appreciably to the relation and the nature of their contribution. It is then dangerous to choose essentially arbitrarily one among a number of different but equally well-fitting relations.

The question of balance between explanatory and pragmatic approaches, i.e. in effect between fundamental and short-term research, in investigating technological problems, whether say in industry or in medicine, raises important and very difficult issues beyond the range of the present discussion. Even in the much narrower context of multiple regression as outlined in the previous paragraph, the distinction between the two approaches is important but not to be taken too rigidly. There is some hope that a prediction equation based on an understanding of the system under study may continue to perform well if the system changes somewhat in the future; any prediction technique chosen on totally empirical grounds is at risk if, say, the interrelationships between the explanatory variables change.

Questions of the specific purpose of the investigation have always to be considered and may indicate that the analysis should be sharply focused on a particular aspect of the system under study, e.g. the proportions outside certain tolerance limits in Examples D and S.

Chapter 2   Some general concepts

2.1 Types of observation

We now discuss briefly some of the types of observation that can be made, by far the most important distinction, however, being that made in Section 1.7 between response variables and explanatory variables.

The first distinction depends on the physical character of the measurements and is between extensive and nonextensive variables. An extensive variable is one which is physically additive in a useful sense: yield of product, count of organisms and length of interval between successive occurrences of some repetitive event are all examples. In all these, regardless of distributional shape, the mean value has a physical interpretation in terms of, for example, the total yield of product from a large number of runs; see Example M connected with the yield of cauliflowers. Thus for extensive response variables, however the analysis is done, the mean value of the variable is among the quantities of interest. Note especially that yield has this property, whereas log yield, or more generally any nonlinear function of yield, does not. An example of a nonextensive variable is blood pressure: the sum of the blood pressures of two individuals has no direct physical interpretation.

The next distinctions depend rather more on the mathematical character of the variable, and in particular on the set of values which it may in principle take. The main possibilities are:

(i) an effectively continuous measurement on a reasonably well-defined scale, i.e. for which a difference of one unit in different parts of the range has in some sense the same interpretation. Analysis will normally be done in terms of the variable itself or some simple function of it;

(ii) an effectively continuous measurement on a relatively ill-defined scale; for example, 'merit' may be scored subjectively on a scale 0 to 100, there being no guarantee that the difference, say 5 to 10, is meaningfully comparable with the difference 80 to 85;

(iii) an integer-valued variable, usually in effect counting numbers of occurrences in some form;

(iv) a discrete variable, often in effect integer-valued, scoring something on an ordered but relatively ill-defined scale, e.g. 4 = very good, 3 = good, 2 = satisfactory, 1 = bad, 0 = very bad. This is broadly equivalent to (ii)

14

with a much reduced range of possibilities. Sometimes the quantitative values, which are essentially conventional, are omitted. Examples N and W illustrate this kind of variable;

(v) a qualitative variable in which the possible values are not ordered, e.g. eye-colour;

(vi) a binary variable, in which there are only two possible forms, e.g. dead, survived; success, failure, etc.; see Examples H, L and X.

Possibilities (i), (iii) and (vi) are the easiest to handle and probably the most widely occurring. In the physical sciences most measurements are of type (i), measuring techniques being well developed, clearly defined and largely standardized. In the social sciences, (iv)–(vi) are relatively common. Any kind of measurement can be reduced to binary form by merging categories, although serious loss of information may be incurred by doing this injudiciously.

2.2 Descriptive and probabilistic methods

A broad distinction has already been made between descriptive statistics, in which no explicit probabilistic element is involved, and methods in which the idea of probability is central. Under descriptive statistics we include the tabulation of data for inspection and the use of graphical techniques. The latter are particularly important, both in the preliminary inspection of data and in the final presentation of conclusions. Current developments in computer graphics may lead to improved ways of dealing with complex relations, especially in several dimensions.

The distinction between descriptive and probabilistically based methods is not a rigid one. Often a probabilistic argument will suggest the calculation of certain quantities which can then be plotted or summarized in a table and regarded as meaningful independently of the original argument which led to their calculation. The method of least squares, which is central to a big part of advanced statistical methods, has various sophisticated probabilistic justifications. It can also very often be regarded as a qualitatively plausible method of fitting. It is, however, a central theme in statistical analysis that important conclusions should have some assessment of uncertainty attached. While sometimes this can be done informally, probabilistic arguments normally play a central role in measuring uncertainty, especially via the calculation of limits of error for unknown parameters.

Most of Part II illustrates methods which have a quite direct probabilistic justification, but it is always important to consider the extent to which the quantities calculated are directly useful as reasonable summaries of the data regardless of the probability model.

The typical form which a probabilistically based analysis takes is as follows. We have observations on one or more response variables and we represent the observations collectively by **y**. Next we consider a family of probability distributions for the observations, i.e. we regard **y** as an observation on a random variable **Y** having a distribution (probability in the discrete case or probability density in the continuous case) usually specified except for values of parameters which are unknown. We call this family of distributions a model. In simple cases the model involves the standard distributions (normal, exponential, Poisson, binomial, etc.) with means (and variances) that depend on any explanatory variables that are available. In the general discussion of this part it aids clarity to distinguish between the observations **y** and the random variable **Y**. In the particular applications of Part II it is much simpler to use **Y** for both random variables and observed values, when it is clear a response variable is under consideration.

Some particular models arise so often in applications that the methods associated with them have been extensively developed. It is, however, very important that the formulation of a model and of scientifically relevant questions about that model are made properly, bearing in mind the unique features of each application; standard models and questions may not be appropriate.

It is useful to have simple examples in mind.

Example 2.1. If the data **y** consist of n repeat observations $(y_1, \ldots, y_n)$ on the same random system, the model will often be that $Y_1, \ldots, Y_n$ are independent and identically distributed random variables. Further, depending on the context, the distribution may be taken as of simple functional form, e.g. normal with unknown mean μ and unknown variance σ^2.

Example 2.2. If the data **y** consist of n pairs of observations (x_i, y_i), where the first variable is an explanatory variable and the second a response variable, it will often be sensible to represent $y_1, \ldots, y_n$ by random variables $Y_1, \ldots, Y_n$ with, for some appropriate function g,

$$E(Y_i) = g(x_i; \beta), \tag{2.1}$$

where $E(Y_i)$ denotes the expected value of Y_i, β being a vector of unknown parameters. An important special case is the simple linear regression model,

$$E(Y_i) = \beta_0 + \beta_1 x_i. \tag{2.2}$$

To complete the model the distribution of the Y's has to be specified. Just one possibility, although a very important one, is that $Y_1, \ldots, Y_n$ are independently normally distributed with the same variance σ^2. A less simple case of Equation (2.1) is Example U in which

$$g(x_i; \beta) = \beta_0 \exp\{\beta_1(x_i - \bar{x})\}.$$

Example 2.3. Equation (2.2) is a special case of a very important general and flexible family of models called univariate normal-theory linear models. These assign the random variables $Y_1, \ldots, Y_n$ independent normal distributions of variance σ^2 and means depending linearly on the explanatory variables with unknown coefficients. That is,

$$Y_i = \sum_{r=0}^{p} x_{ir}\beta_r + \varepsilon_i \qquad (i = 1, \ldots, n), \qquad (2.3)$$

where ε_i is an unobservable random variable representing error, $\beta_0, \ldots, \beta_p$ are unknown parameters, and x_{ir} is the value of the rth explanatory variable corresponding to the ith observation y_i. The random errors ε_i are independently normally distributed with zero mean and variance σ^2. This model allows the inclusion of simultaneous dependence on several explanatory variables. Very often $x_{i0} = 1$ $(i = 1, \ldots, n)$ corresponding to a 'constant' term β_0 in the relation.

2.3 Some aspects of probability models

The following comments about models are of general importance and apply to virtually all the subsequent discussion.

(i) There are at least two different physical meanings possible for the model. In the first there is a well-defined population of individuals, and the individuals observed (the sample) are drawn from the population in a 'random' way. The probability model then specifies the properties of the full population. Here the population is real and could in principle be fully measured. A more common situation is where observations are made on some system subject to random fluctuations and the probability distributions in the model specify what would happen if, entirely hypothetically, observations were repeated again and again under the same conditions. Sometimes the word 'population' is used again in such contexts but the population is now a hypothetical one. In some contexts the whole idea of repetitions is notional. For example, a probability model of monthly unemployment numbers for a particular country essentially amounts to saying that it may be useful to treat the numbers, which are of course unique, as if they were generated by a physical random mechanism capable of repetition under the same conditions. Example C, concerned with literary data, is another instance where the relevance of a probability model is indirect.

(ii) The distributions arising in the models nearly always involve unknown parameters. These play a crucial role. Much of statistical theory is taken up with the question of how to use the data as effectively as possible to answer questions about unknown parameters. For any particular part of the analysis we distinguish between the parameters of interest, i.e. under direct study, and the others which we call nuisance parameters. In Example 2.1 above, μ might be the parameter of interest and σ^2 a nuisance parameter, especially when the variable is extensive or the random variation is error rather than natural variation. When the data are sampled from existing populations the parameters describe properties of the populations and it is natural to be interested in them. In other cases we normally think of the parameters as constants determining the nature of the random systems under investigation free of the accidental particular random disturbances encountered in the data. Parameters representing fundamental physical constants, as in Example O, are an illustration. Also the parameters and probability distributions in effect determine the distribution of future observations that might be obtained on the same random system in the future. Occasionally we work explicitly with an answer expressed in terms of predicted future observations rather than with parameters; see again Examples D and S.

(iii) The model is always tentative. In some contexts it is pretty clear from previous experience what a suitable family of models is likely to be, and rough inspection of the data for gross discrepancies may be all that is necessary to examine the adequacy of the model. In other cases, especially with rather complex data from an unfamiliar field, it may be far from clear what is the best formulation. A very important part of the analysis is then a preliminary analysis to search for a suitable model. This involves not just testing the adequacy of any initial model, but doing so in a way that will suggest better models and bring to light possibly unsuspected effects in the data.

(iv) Quite often it is useful to distinguish two aspects of a model. The primary aspect serves in effect to specify the main questions of interest. The secondary aspects serve to complete the model and thus to indicate the analysis likely to be suitable, and the precision of the conclusions. For instance, in the simple linear regression model of Example 2.2 above, interest would usually, although not necessarily, be concentrated on the dependence of $E(Y)$ on x and in particular on the parameter β_1, which specifies the change in $E(Y)$ per unit change in x. Assumptions of normality, etc., would then be secondary, not in the sense of being unimportant, but rather in affecting the conclusion indirectly, whereas a wrong specification of the dependence of $E(Y)$ on x leads to an ill-conceived question being considered. The primary aspect is more important in the sense that an approximate answer to the 'right' question is to be preferred to a very precise answer to a 'wrong' question. The distinction between primary and secondary aspects is a subject-matter issue.

(v) Most models involve at some point an assumption that certain random variables are mutually independent and this is often one of the most sensitive assumptions made. While such an assumption can to some extent be tested from the data, consideration of the way the data are obtained is usually the main guide as to what independence assumptions are reasonable. Observations obtained on the same individual at nearby time points will not normally be independent. The rather rare exception is when pure error of measurement is the dominant source of variation. Another example is that in Example 2.2 above, the n points (x_i, y_i) might correspond to $n/2$ individuals each observed twice. To be quite specific, measurements on x and y might be taken on the left eye and on the right eye of a number of men. It would normally be very misleading to treat the random variables Y_i as mutually independent; in extreme cases they might be nearly equal in pairs. The assumption that Y_i corresponding to different men are independent would, however, quite often be made. Specification of independence is one of the most commonly occurring sources of difficulty in choosing a model. The more complex forms of analysis of variance, such as Examples I, Q and R, illustrate the care needed in dealing with structured random variation.

Chapter 3   Some strategical aspects

3.1 Introduction

In the previous chapter we introduced some of the general ideas that are involved in statistical analysis and in the following chapter we outline the main kinds of technique that are useful. The present chapter develops some of the broader strategical ideas involved in applying statistical techniques. These ideas are of greatest importance in complex investigations and therefore are relatively difficult to illustrate in the small-scale examples in Part II of this book. They bear also on the interplay between statistical analysis and subject-matter interpretation.

3.2 Incorporation of related data and external information

The scientific or technological interpretation of investigations never proceeds in isolation, although individual investigations may have very considerable, even if ultimately temporary, importance. In each situation some reasonable balance has to be drawn between, on the one hand, making individual investigations satisfactory in their own right, and, on the other hand, achieving a synthesis of information from many sources. The latter requires the fitting of forms of systematic relation that will apply widely, that are consistent with theoretical information and any known limiting behaviour, and which will allow testing of the consistency of separate sources of information.

In principle, one should aim for description of all available data in a single form of model with as many parameters as possible constant over all data sets, and variation that does occur between data sets should, so far as is feasible, be 'explained'. The essentially iterative nature of this is important. The phases of preliminary and definitive analysis are not clearly isolated.

Often it will be realized that earlier data have been inappropriately analysed; it will be for consideration whether it is worth re-analysing them.

Where previous work, general knowledge of the specific situation, or reasonably well accepted theory, suggest, say, a particular form of regression relation, it will usually be wise to base the initial analysis on that model, possibly supplemented by additional parameters to allow adequacy of fit to be examined. If a clear discrepancy with previous work is isolated, it should be given a rational interpretation and, at least in principle, a new formulation should be found that explains the previous and current data in a single setting.

3.3 Role of special stochastic models

A relatively recent emphasis in statistical work has been on special theoretical stochastic models representing simplified theories of the system under investigation. Their properties can be studied either mathematically or by computer simulation. The use of these in statistical analysis is an attempt to relate data to some underlying physical or biological mechanism and therefore some of the points in Section 3.2 are relevant. Despite its great interest, some of the work on stochastic models seems rather remote from data. Often to be at all realistic, models have to be very complicated and involve many parameters, so that discrimination between alternative models requires extensive data of high quality.

The use of simple models for qualitative understanding can, of course, be very fruitful. Quite possibly, more use of very simple models could be made in the analysis of data, one focal point of the analysis being the careful description of the ways in which the model does not fit. Thus comparison with a Poisson process is a natural way to describe systems of point events; see Examples A and T. Certainly there is a need in planning analyses to bring empirical analysis of data and theoretical analysis of stochastic models into closer contact, and concern with the development of special stochastic models is likely increasingly to become a feature of statistical work.

3.4 Achievement of economical and consistent description

In relatively complicated problems, as typified by regression problems with many explanatory variables, the main objective is an economical description of the data in a form consistent with external information. There are two broad approaches, called 'forward' and 'backward'.

The 'forward' approach is to start from a relatively simple point and to complicate the model only when the data explicitly indicate the need to do so; if the starting point is soundly based, and not an arbitrary choice among several radically different possibilities, the use of the 'forward' approach is an expression of mildly sceptical optimism. The backward' approach is to start from a complicated model and to simplify it as much as the data permit.

In principle, the 'backward' approach starting with a rich family of models and ending with a list of all those simple models reasonably consistent with the data, bypasses the difficulties associated with complex sequences of inter-related choices; see Example C on 'clustering' of literary works. Any choice between alternative quite different models fitting roughly equally well has to be made explicitly on grounds external to the data.

The 'backward' approach is the safer one and should normally be used when it is not too ponderous and especially when there is a major interest in and uncertainty over the primary formulation of the problem. The 'forward' approach is more appropriate for the secondary aspects of the problem, e.g. over the structure of error. In any case it is obviously impossible and undesirable to start with a model that covers all possibilities that could arise, so that some mixture of the two kinds of approach is nearly always inevitable. Examples G, R and W illustrate in simple form the arguments involved.

For example, in the classical orthogonal analyses of variance a version of the 'backward' approach is to start with the full analysis of variance indicated by the logic of the design, and to hope that this will point to a simple description, e.g. one with many interactions negligible. A more extreme version of the 'backward' approach would allow from the beginning the possibility of transformation of the response variable. A version of the 'forward' approach is to start with a model involving, say, only main effects and to add to it as seems necessary; see Example P on the analysis of a 'random balance' experiment. In more conventional sets of balanced data the 'backward' approach seems normally much preferable. It is simple, yet allows the detection of unanticipated complexities and also the possibility of finding simple structure other than that tied to main effects and low order interactions.

Another aspect of the 'forward'–'backward' contrast, already touched on, concerns the analysis of data in rational sections. For instance, with data on men and women, one may:

(i) first ignore sex differences and then check residuals for the possible presence of sex differences;

(ii) fit a model with a single parameter for a sex difference and then check the residuals for interactions;

(iii) fit a model allowing for a sex difference and for some sex × explanatory variable interactions;

(iv) analyse the data for men and women separately, aiming for a single model to be formulated after inspection of the results of the separate analyses.

Form (iv) can be regarded as a version of the 'backward' approach. In very complex problems there may be a number of advantages, e.g. ease of communication, in analysing the data in manageable sections. Procedure (iii) can be useful, as in Example G, if a section is too small for separate analysis.

The issues touched on here are central ones in the strategy of statistical work. The ability to go 'forward' straight to the simple essence of a complex topic is to be highly prized. The 'backward' approach is likely often to be slower, but perhaps safer, provided that the complex starting point is judiciously chosen and regarded as no more than a basis for ultimate simplification. The task is usually to judge what mixture is appropriate.

With very extensive data it will frequently be good to begin by analysing suitably small subsections of data. A very widely applicable approach is to use an analysis of subsections of data to replace each subsection by a small number of summarizing quantities which can be regarded as derived response variables for a second stage of analysis. Even, however, where a single unified analysis is aimed at, as for example in applying analysis of variance to relatively complicated but balanced multifactor systems, it can be very helpful to start from the analysis of easily manageable subsections of data; see, for instance, Examples K and S.

A different but related aspect is the question arising widely in economic statistics and econometrics, and more broadly in the analysis of official statistics, of the merits of aggregation and disaggregation. Simplicity calls for aggregation, i.e. the merging of similar and related sets of data, whereas investigation of detailed effects may call for disaggregation, i.e. breaking down of data into component parts. Modern capabilities of fitting models with quite large numbers of parameters allow a greater degree of disaggregation than was formerly possible but it is hard to make useful comments in generality. Ideally data should be available in a form for checking the absence of interaction (see Section 4.13) one level higher than that used in the final presentation of conclusions.

3.5 Attitudes to assumptions

The broad division mentioned in Section 3.4 has some parallel in attitudes to the assumptions underlying the more formal probabilistically based statistical methods. The additional power and subtlety of probabilistically based methods are bought at a price, namely the introduction of assumptions about the probability distribution supposed to have generated the data. These assumptions are virtually always idealized representations of the real situation and the question is not so much whether the assumptions are exactly correct as to:

(i) whether it is possible from the data to obtain a clear indication of how to improve the model;
(ii) whether it is likely to be important to do so; and
(iii) whether critical assumptions in the model can be bypassed.

One important approach, essentially that common in applied mathematics generally, is to be very critical about assumptions judged central to the problem, but to make quite drastic simplifying assumptions about secondary aspects; at the end of the analysis one considers, at least qualitatively, how sensitive the main conclusions are to the assumptions made. This is essentially the spirit underlying the use of such methods, to be outlined later, as least squares and normal-theory tests of significance, preceded or followed by inspection for anomalous points. This is one facet of the 'forward' approach. One corresponding 'backward' approach is to make minimal reasonable assumptions feasible without major loss of efficiency. In relatively simple situations, and where cautious testing of significance is of central importance, use of nonparametric methods may be wise. In more complex situations where concise description of effects is required, and tests of significance play a subsidiary role, the use of simplifying assumptions about, for example, error structure will usually be needed. Drastic simplification of aspects subsidiary to the main purpose is essential in relatively complex problems and elaborations of the secondary assumptions that have little effect on the final conclusions should, wherever feasible, be avoided. An illustration when the final description is in terms of a model which does not quite fit is Example T on the analysis of intervals between equipment failures.

3.6　Depth and complexity of analysis appropriate

Judgement of the amount of detail sensible in analysis is difficult. There are a number of aspects. What should be the relative effort devoted to data collection and to analysis, especially in fields where relatively large amounts of data can be obtained fairly easily? The efforts should not be too disparate. When is it sensible to try to settle an outstanding issue by more refined analysis of data already available rather than by getting new observations? How elaborate a preliminary analysis is desirable?

In the central analysis, the wide availability of computer packages has made quite complicated methods of analysis painlessly available even to those with little statistical training, and the dangers of this have often been discussed. There is another side, however; it is now also painless for the statistician to generate large numbers of plots of residuals, tests of normality, analyses on numerous different scales, etc., and thus to get very large amounts of computer output even from very simple data. Without some ruthlessness in appreciating that the great majority of this must be discarded, the objective of achieving and reporting simple conclusions is threatened.

A particularly difficult issue is to decide when simple descriptive analysis is adequate and when explicit calculations of uncertainty involving probabilistic arguments are called for. If straightforward descriptive methods, graphical or numerical, lead to apparently clear-cut conclusions, the role, if any, of probabilistic methods is in danger of being confined to attaching a somewhat objective seal of approval. The necessity or otherwise of this depends on the circumstances, although pressures to apply, say, tests of significance to totally obvious effects should be resisted, except perhaps as a training exercise. Note, however, that calculation of limits of error for the size of effects may still be valuable. A more interesting application of probabilistic ideas is to the isolation, description and confirmation of effects in relatively complex situations, although even here the possibility of simpler, largely descriptive, methods for the presentation of conclusions is important.

To pursue the same general point, if in relating a response variable y to an explanatory variable x, a plot shows a very simple, e.g. virtually linear, relation, further analysis may be superfluous; see Example B. If, however, it is required either now or in the future, to compare the slope with the slope from similar data obtained under rather different conditions, or if it is required to put limits of error on a prediction from the line, then probabilistic methods will find fruitful application.

3.7 Analysis in the light of the data

An important strategical issue concerns the extent to which methods of analysis are to be settled in advance and the extent to which it is legitimate and desirable to modify the approach to final analysis in the light of preliminary analysis.

Firstly, it is highly desirable in the planning stages of an investigation to consider at least in broad outline how the analysis is to proceed. Even further, it is normally wise to try to anticipate the main qualitative kinds of response pattern that might be observed, to check that the information for their proper interpretation will be available. In major investigations, especially ones involving collaboration between different centres and where the conclusions may be controversial, it is desirable to go still further and in advance to set out a scheme of analysis in some detail. This applies, for example, to some major clinical trials, and to large-scale international studies of cloud seeding.

Yet, especially with complex data, it is quite unrealistic to suppose that the form of appropriate models and questions can be precisely settled in advance. Further, there is always the possibility that the data will show some entirely unanticipated features of considerable importance; to ignore such features just because they were not part of the scheme of analysis set out in advance would be foolish. Moreover, in initial investigations of fields about which little is known in advance, bodies of data may have to be analysed with not much idea to begin with of what aspects are likely to be interesting.

For these reasons, preliminary analysis will often, although by no means always, be needed and the possibility of modification of analysis in the light of the data always kept open.

In studies where the protocol sets out a detailed method of analysis, it is clear that the agreed method of analysis should always be carried through and reported. If the data show that analysis to be inappropriate, the reasons for this then need to be argued and the preferred analysis defended.

In other cases, the effect of modifying analysis in the light of the data is different depending on whether the modification:

(i) affects only the secondary aspects of the model;

(ii) changes the precise formulation of the primary aspects of the model, although leaving the central question unchanged;

(iii) leads to a new focus of interest.

For example, suppose that primary interest lies in the slope of a linear regression relation. Unexpectedly the data suggest that the relation, while linear, is such that the variance of response, Y, increases markedly with the explanatory variable, x. Ordinary least squares is therefore replaced by weighted least squares. So long as the change in variance is of no intrinsic interest, we have situation (i). Suppose now that in a similar situation it is found that linear regression of log Y on log x would give a better linear fit than Y on x. The qualitative question, namely the study of a monotonic relation between Y and x has not changed, although its precise mathematical formulation has. This is situation (ii). In both cases it does not seem that any special adjustment to the final analysis is called for to allow for choice in the light of the data.

Suppose, however, that in the first example the change in variance is of intrinsic interest and that in particular it is required to assess the significance of the change. A test of the null hypothesis that the variance is constant can be set up, but testing that hypothesis has been suggested by the magnitude of the departure observed in the data. It is clear that direct use of the ordinary significance level may overestimate the evidence in favour of the effect under study. The usual procedure is to introduce an adjustment for selection. This is approximately that if the effect studied is the largest of m effects that might have been chosen for study, the observed significance level is to be multiplied by m. This maintains at least approximately the physical interpretation of a significance level (see Section 4.7) as the probability of a deviation as or more extreme than that observed arising by chance. In some cases the value of m may be unclear, although of course in extreme cases it may not be necessary to specify m with any precision. Thus possibility (iii) of those listed above is the one where allowance for adjustment of the analysis in the light of the data may be crucial.

A conceptually different but formally rather similar situation would arise if for a null hypothesis specified in advance one were to apply several tests, e.g. parametric tests under rather different detailed assumptions and nonparametric tests of different kinds, and then to choose the test giving the most significant result. This is a poor approach, dishonest if what is done is concealed; if it were to be adopted, however, some adjustment for selection would again be essential.

To summarize, the one case where special adjustment is needed for adopting an analysis chosen in the light of the data is where the statistical significance of an effect suggested by the data is at issue. The matter is not totally a technical statistical issue. In some fields it may be feasible to take the line that effects whose study was not the original objective need an independent confirmatory investigation before their existence can be regarded as established. Indeed, in the long term this is a sensibly cautious approach, but in subjects where data take a long time to obtain some assessment of significance in the initial investigation, even if its interest is ultimately temporary, can still be valuable.

In this, as indeed in most other stages of statistical analysis, fruitful interplay between statistical and subject-matter considerations is of critical importance.

Chapter 4 — Some types of statistical procedure

4.1 Introduction

The objective of statistical analysis is to discover what conclusions can be drawn from data and to present these conclusions in as simple and lucid a form as is consistent with accuracy. As outlined in Section 1.4, there will typically be a number of phases to the analysis. Within the phases of preliminary and definitive analysis, technical devices of various kinds are quite commonly used, and these we now outline.

Understanding of these ideas is achieved partly by experience with applications and the following very brief account is meant as a general guide, to be read in conjunction with the applications in Part II, and with more systematic discussions.

4.2 Formulation of models: generalities

The more formal statistical procedures are based on a probability model for the data and we begin with some general comments on model formulation.

The setting up of such a model for a particular application can be a crucial issue and, as with problems of mathematical formulation in general, it is hard to give more than general guidelines on how to proceed. It will be an advantage to use wherever reasonably possible a fairly standard model, or minor modifications thereof, so that methods of analysis, including appropriate computer programs, are readily available.

Models are always to some extent tentative. This is obviously so in preliminary analysis, but even in what is hoped to be a definitive analysis formal or informal checks on model adequacy are required.

We have already in Section 2.3 (iv) outlined an important distinction between the primary aspects of a model, which in effect serve to define the problem under study, and the secondary aspects, which complete the specification in sufficient detail to allow the calculations of the precision of the final conclusions and, more broadly, which indicate appropriate methods of analysis.

Very often another distinction can usefully be drawn, namely between the model for the systematic part of the variation and the model for the haphazard part. This distinction is not always clear cut, but the general idea will usually be fairly clear. Thus in Example 2.2 the systematic relation between a response variable Y_i on the ith individual and an explanatory variable x_i is such that

$$E(Y_i) = g(x_i; \beta), \tag{4.1}$$

where g is a known function and β is a vector of unknown parameters.

In some applications the distribution of Y_i may be determined by the mean, as for example in Poisson, binomial and exponential distributions. In other applications it may be possible to express Y_i ($i = 1, \ldots, n$) in terms of a set of independent and identically distributed random variable $\varepsilon_1, \ldots, \varepsilon_n$, say in an extended notation

$$Y_i = g(x_i, \varepsilon_i; \beta). \tag{4.2}$$

For instance, we may have

$$Y_i = g(x_i; \beta) + \varepsilon_i, \tag{4.3}$$

where the ε_i are independent and identically distributed random variables of zero mean.

In these cases, modelling of the systematic and random components of the variables can be considered separately. Note, however, that if the shape of the distribution of Y_i changed notably with the explanatory variable, such separation of systematic and random components of the system might not be feasible. If the random variation represents 'error' rather than 'natural' variation, the systematic part of the model describes the variation that would be observed were error eliminated.

The contrasts 'primary' versus 'secondary' and 'systematic' versus 'random' are logically quite separate. While it is quite conceivable that the primary aspects could include or consist of the random part of the model, this is in practice relatively rare. Where the effect of the explanatory variables can be entirely captured within the systematic part of the model, it is likely that the primary aspect of the model will be all or part of the systematic component.

4.3 Formulation of models: systematic component

It is clearly impracticable to give a comprehensive discussion of all the mathematical functions that might be used to describe systematic variation. Here we discuss briefly some of the considerations that enter into such a choice, concentrating on the description of how the expected value of a response variable, $\eta = E(Y)$, depends on an explanatory variable x or more generally on several explanatory variables $x_1, \ldots, x_p$.

Firstly it is often worth distinguishing cases in which the relation $\eta = \eta(x)$:

(i) is adequately described by specifying, numerically or graphically, values of η for suitable chosen values of x, without fitting a mathematical formula.

Thus in Example B mathematical description of the smooth variation seems superfluous;

(ii) is to be fitted by a mathematical formula which is to be used for interpolation or numerical differentiation and integration, or for future prediction, but in which no detailed scientific interpretation is to be based on the precise form of equation to be used or on the parameters in the assumed relation;

(iii) is to be fitted by a mathematical formula, the particular form being the basis of at least a qualitative interpretation, including often the use of particular parameters to describe concisely the difference between different similar sets of data.

The second case is conceptually easier than the third. To an appreciable extent the particular mathematical form used is secondary in (ii) and the main matter for consideration is the extent of smoothing that is appropriate in fitting. Computational simplicity suggests fitting a model linear in unknown parameters, flexible families of such models being polynomials and Fourier series. Alternatively, empirical smoothing techniques not based on explicit formulae for $\eta = \eta(x)$ can be used. If, however, some extrapolation is required, or if the data are rather poor in some parts of the x-range of interest, more attention to careful choice of a model may be wise and some of the following considerations become relevant.

We now concentrate on (iii), where choice of an appropriate model is more critical. The following general points arise.

(a) If some limiting or boundary behaviour is known it will usually be wise to fit a model consistent with that behaviour. For example, suppose that when $x = 0$, $E(Y) = 0$, i.e. $\eta(0) = 0$. Now if the data are obtained for values of x fairly remote from $x = 0$ it is entirely conceivable that the data are consistent with a straight-line relationship not through the origin. If, however, they are consistent also with $\eta(x) = \beta_0 x^{\beta_1} (\beta_1 > 0)$, the latter is to be preferred as being consistent with the boundary behaviour. Similarly, gently curved data observed over a relatively narrow range may be consistent with a parabola, $\eta(x) = \beta_0 + \beta_1 x + \beta_2 x^2$, but if general considerations suggest that $\eta(x)$ tends to a limit as $x \to \infty$, some other function should be fitted for any but very restricted purposes. See also Example U on the analysis of survival times.

(b) If there is a theoretical treatment of the system under study, a model that establishes a link with that theory will be valuable. In a 'forward' approach we start from the theoretical model and amend it if and only if the data so dictate. In a 'backward' approach we extend the theoretical model by adding additional parameters and recover the theoretical solution as one special case, if the data so allow. For example, if we started with the theoretical model

$$\eta(x) = \beta_0 + \beta_1 e^{-\beta_2 x}, \tag{4.4}$$

we could add as additional terms polynomials in x, e.g. a term $\beta_3 x$. An alternative and in some ways preferable possibility is to add terms in the partial derivatives of η with respect to the β_r, leading here to the model

$$\eta(x) = \beta_0 + \beta_1 e^{-\beta_2 x} + \beta_3 x e^{-\beta_2 x}. \tag{4.5}$$

Equation (4.5) has the advantage of being invariant under transformations, e.g. of x into $\log x$.

(c) Previous data may indicate an appropriate form at least for preliminary analysis. Consistency with previous data should wherever feasible be checked.

(d) The ability to fit a range of similar data is of great importance in model choice. It will usually be wise to aim to fit all the sets of data with the same form of model. For instance, if in a number of sets of similar data, some are adequately fitted by straight lines whereas others need to be fitted by parabolas, it will normally be sensible to fit all the sets of data by parabolas, unless some rational explanation can be found for the split into two kinds.

(e) Low-degree polynomials can be valuable for interpretation as well as for fitting for interpolation. In particular, in a parabola

$$\eta(x) = \beta_0 + \beta_1 x + \beta_2 x^2, \tag{4.6}$$

the parameter β_2 gives an indication of the direction and amount of curvature; see Example F on the analysis of process effects in the presence of trend. A cubic is occasionally useful for indicating a point of inflexion, but higher-degree polynomials by and large are not very helpful for interpretation, partly because of the difficulties of interpreting the separate parameters.

Tables 4.1 and 4.2 give a number of special relations useful in fitting systematic relations. These may be used empirically, i.e. a choice of relation may be based solely on inspection of the data and on the adequacy of the fit. Alternatively, a relation may be suggested by some considerations external to the data, for example by study of a deterministic or stochastic theory of the system. The linearized form of the relation, where it exists, is often useful both for initial fitting and for examining adequacy of fit.

For problems with more than one explanatory variable, an alternative to the fitting of a relation to all explanatory variables simultaneously is to fit in stages. For instance, with two explanatory variables x_1 and x_2, it may be possible to divide the data into sets with exactly or nearly constant x_2 and then to fit a relation of the same form within each set for the variation with x_1. Then the parameters, or rather their estimates, can be related to x_2 and hence a single model constructed.

The above discussion is for quantitative explanatory variables. If the explanatory variables are qualitative, such as qualitatively different treatments, we have to describe how a suitable aspect of response, for instance a

Table 4.1. A few formulae useful in describing systematic variation: one explanatory variable

		Linearized form*
1. Line through origin	$\eta = \beta x$	—
2. Line	$\eta = \beta_0 + \beta_1 x$	—
3. Second degree (parabola)	$\eta = \beta_0 + \beta_1 x + \beta_2 x^2$	—
4. General polynomial	$\eta = \beta_0 + \beta_1 x + \ldots + \beta_k x^k$	—
5. Power law	$\eta = \beta_0 x^{\beta_1}$	$\log \eta = \gamma_0 + \gamma_1 \log x$
6. Power law (non-zero origin)	$\eta = \beta_0 (x - \delta)^{\beta_1}$ $(x > \delta)$	$\log \eta = \gamma_0 + \gamma_1 \log(x - \delta)$
7. Exponential decay or growth with zero asymptote	$\eta = \beta_0 e^{-\beta_1 x}$	$\log \eta = \gamma_0 + \gamma_1 x$
8. Exponential decay or growth	$\eta = \beta_0 e^{-\beta_1 x} + \beta_2$	$\log(\eta - \beta_2) = \gamma_0 + \gamma_1 x$
9. Mixture of exponentials	$\eta = \beta_{01} e^{-\beta_{11} x} + \beta_{02} e^{-\beta_{12} x} + \beta_2$	—
10. Passage between two asymptotes	$\eta = \beta_0 + \dfrac{\beta_1 e^{\beta_2 (x - \beta_3)}}{1 + e^{\beta_2 (x - \beta_3)}}$	$\log \left(\dfrac{\eta - \beta_0}{\beta_0 + \beta_1 - \eta} \right) = \gamma_0 + \gamma_1 x$
11. Modified exponential decay or growth	$\eta = \beta_0 \exp(-\beta_1 x^{\beta_2})$ $(x \geqslant 0)$	$\log\{-\log(\eta/\beta_0)\} = \gamma_0 + \gamma_1 \log x$
12. Periodic wavelength $2\pi/\omega_0$	$\eta = \beta_0 + \beta_1 \cos(\omega_0 x) + \beta_2 \sin(\omega_0 x)$	—

* The linearizing transformations are stated for positive β's and may need modifications of sign otherwise. The β's and γ's are adjustable parameters, different for each case. The γ's are functions of β's. All equations can be reparameterized.

mean, varies between levels. With several different qualitative explanatory variables, it will be usual to employ the concepts of main effects and inter-actions to attempt to simplify the description; for balanced data, the elegant techniques of analysis of variance lead to systematic investigation of those interrelationships that can be studied from a given design. See Examples K and L. The same ideas can be applied very widely although their implementa-tion can be computationally difficult in situations where nonlinear methods of analysis must be employed.

For mixtures of qualitative and quantitative explanatory variables a choice between describing how the effect of the qualitative variables depends on the quantitative variables, or how the response pattern to the quantitative variables depends on the qualitative variables, will depend on the context.

Table 4.2. A few formulae useful in describing systematic variation: p explanatory variables

		Linearized form
1. Plane through origin	$\eta = \beta_1 x_1 + \ldots + \beta_p x_p$	—
2. Plane	$\eta = \beta_0 + \beta_1 x_1 + \ldots + \beta_p x_p$	—
3. Second-degree surface	$\eta = \beta_0 + \beta_1 x_1 + \ldots + \beta_p x_p$ $+ \beta_{11} x_1^2 + \ldots + \beta_{pp} x_p^2$ $+ 2\beta_{12} x_1 x_2 + \ldots + 2\beta_{p-1,p} x_{p-1} x_p$	—
4. General polynomial	—	—
5. Power law	$\eta = \beta_0 x_1^{\beta_1} \ldots x_p^{\beta_p}$	$\log \eta = \gamma_0 + \gamma_1 \log x_1 + \ldots$ $+ \gamma_p \log x_p$
6. Power law (nonzero origin)	$\eta = \beta_0 (x_1 - \delta_1)^{\beta_1} \ldots (x_p - \delta_p)^{\beta_p}$ $(x_1 \geqslant \delta_1, \ldots, x_p \geqslant \delta_p)$	$\log \eta = \gamma_0 + \gamma_1 \log(x_1 - \delta_1) + \ldots$ $+ \gamma_p \log(x_p - \delta_p)$
7. Exponential decay or growth in all variables, zero asymptote	$\eta = \beta_0 \exp(-\beta_1 x_1 - \ldots - \beta_p x_p)$	$\log \eta = \gamma_0 + \gamma_1 x_1 + \ldots + \gamma_p x_p$
8. Exponential decay or growth	$\eta = \beta_0 \exp(-\beta_1 x_1 - \ldots - \beta_p x_p)$ $+ \beta_{p+1}$	$\log(\eta - \beta_{p+1}) = \gamma_0 + \gamma_1 x_1 + \ldots$ $+ \gamma_p x_p$

The arrangement and conventions are given in the footnote to Table 4.1.

4.4 Formulation of models: random component

The formulation of a suitable model for the random component of variability is much more a technical statistical matter than is the discussion in Section 4.3 of systematic variation. From a general point of view there are typically three broad matters to be considered:

(i) what assumptions of independence are appropriate in setting up a model for the random variation;

(ii) what formulation will lead to a type of error distribution as stable as possible over the whole data;

(iii) what assumptions are appropriate about the form of the probability distribution of the random variation.

Of these, (i) is frequently the most critical aspect. In a carefully designed experiment or sample survey, the structure of the randomization in the design will indicate the appropriate independence assumptions and a hierarchy of error structure may apply, to be investigated by careful use of analysis of variance, see Examples Q, R and S. In observational studies, however, the reasonable assumptions to make may be less clear. Frequently observations taken close together in space or time, or by the same observer or on the same set of apparatus, are to be treated as positively correlated; see some of the discussion of the retrospective study, Example V. Common sources of mistake, very easily made unless the method of obtaining the data is quite clearly understood, are as follows:

(a) to treat n observations as independent when in fact they consist of, say, r repeat observations on m individuals, $rm = n$;

(b) to treat as independent repeat observations made by the same observer close together in time, the observer knowing that the observations 'ought' to be the same; and

(c) to ignore serial correlations in studies of a phenomenon varying in time.

One characteristic of statistical studies should be a careful study of the structure of random variation. An uncritical assumption that random variation is modelled by independent and identically distributed random variables is frequently naive, although of course sometimes justified.

In connection with point (ii), the use of as stable a form of error distribution as possible, it will be central, especially if the random variation is the secondary aspect of the model, that random variables of essentially the same form are used for the whole data. Thus we might describe each stratum of error variation by normally distributed random variables of constant variance. In some contexts an error variance might be different in different sections of data, but a model in which quite different assumptions are made in different parts of the data should be avoided if feasible. See Example T on the comparison of failure rates.

Over point (iii), the form of probability distribution to use, the nature of the problem may indicate one of the fairly standard distributions, normal, exponential, gamma, Poisson, binomial, and so on, as a reasonable starting point, possibly after transformation of the response. If distributional shape is of primary interest, attention will focus on a test of distributional form, possibly by comparison of a nonparametric estimate of distributional shape with the theoretical form, and possibly by the calculation of suitable test statistics for distributional shape. If distributional shape is a secondary aspect it will often be simple to begin with a special simplifying assumption and then later to consider whether that assumption has materially affected the conclusions. If very careful and cautious testing of significance is required, a nonparametric or distribution-free analysis may be sensible, usually assuming that the error variation is independent and identically distributed with unspecified form. Note that, of course, such an analysis does not avoid assumptions, especially ones of independence, about the random variation present. A further serious disadvantage of nonparametric methods, when error structure is a secondary aspect, is that the analysis of complex dependencies tends to be cumbersome.

4.5 Calculation of summarizing quantities

It is often required to replace sets of data, e.g. groups of observations obtained on a particular batch of material, by one or more summarizing statistics describing aspects of the data, for instance, position, scatter, or trend. These summarizing statistics may be used:

(i) as a final summary analysis, conclusions being drawn by inspection of tables;

(ii) as an intermediate stage in analysis, the final analysis being an examination of how the summary statistics, or derived variables, depend upon explanatory variables;

(iii) as a basis for graphical analysis.

We discuss graphical methods separately in a moment.

Often summarizing statistics are chosen from purely descriptive considerations, as in the literary study, Example C, and in Example E. If, however, their choice is to be approached probabilistically, it can usually be regarded as a problem in point estimation. There is a particular component of the unknown parameter vector that is of interest. From the relevant data vector y we calculate a function $t = t(\mathbf{y})$ intended to be as close to the parameter θ of interest as possible. To the quantity t, often called a point estimate, there corresponds a random variable T having for each possible parameter value θ, a probability distribution with density, say $f_T(t;\theta)$. This distribution is usually called the sampling distribution of T. It will in general depend on nuisance parameters, i.e. on unknown parameters not of immediate interest in their own right. Its physical meaning is that it gives the hypothetical frequency distribution of the estimate t if repeat sets of data were obtained under the same conditions and t calculated from each set. We now want to choose a function $t(\mathbf{y})$ such that $f_T(t;\theta)$ is as closely concentrated around θ as possible, whatever may be the true θ. Once this is expressed more formally it is a mathematical problem to find a suitable t, subject of course to feasibility of computation.

One very particular way of expressing the requirement is that we should insist that

$$E(T) = \theta \tag{4.7}$$

and subject to this choose T so that

$$\text{var}(T) = E(T - \theta)^2 \tag{4.8}$$

is a minimum. Estimates satisfying (4.7) are called unbiased. Estimates satisfying both Equations (4.7) and (4.8) are called minimum-variance unbiased.

4.6 Graphical analysis

Graphical methods are important both in the preliminary analysis of data and in the final presentation of conclusions. There will be many illustrations in Part II and here we outline briefly a few general points, although it is difficult to give a systematic discussion.

It is relatively easy to detect by eye systematic departure from linearity and, with a little training in necessary caution, to detect departure from a totally random plot. This suggests that we should arrange plots so that under 'ideal' conditions either a straight line or a random plot should result. The former is most relevant when the systematic part of the variation is the focus of attention and the latter when the random components, e.g. the residuals from a fitted linear model are under study; see, for instance, the regression analysis of Example G and the exponential plots in Example T.

It is also an aid to interpretation, although not always achievable, to arrange that points have independent errors, preferably of equal and known standard errors; this eases assessment of the reliability of the plot. Of course the non-linear transformations that may be necessary to induce approximately linear plots may be in conflict with these other requirements. Note, however, that it is usually unwise to adopt a nonlinear transformation to linearize a plot if thereby ranges of a variable of little interest are accentuated. For example, a transformation from survival time to log survival time would have the effect of accentuating the behaviour near zero survival time and contracting the part of the graph connected with long survival times and, depending entirely on the context, this might be undesirable; see again Example T.

When graphical methods are used in preliminary analysis the precise form of arrangement used is not critically important. For the final presentation of conclusions, however, careful attention to format is desirable and the following guidelines are suggested:

(i) axes should be clearly labelled with the names of the variables and the units of measurement;

(ii) scale breaks should always be used for 'false' origins;

(iii) comparison of related diagrams should be helped, for example by using identical scales for plotting and by placing related diagrams on the same or facing pages;

(iv) scales should be arranged so that systematic and approximately linear relations are plotted at roughly 45° to the coordinate axes;

(v) legends should make diagrams as nearly self-explanatory, i.e. independent of the text, as is feasible;

(vi) interpretation should not be prejudiced by the technique of presentation, for example by superimposing thick black smooth curves on largely random scatters of faint points;

(vii) too much information should not be put in one graph, either by putting too many points or by supplying unduly extensive supplementary information.

To some extent the choice between graphical and tabular display is a matter of taste. Graphical methods are on the whole the more suitable for showing broad qualitative features. Tabular methods are definitely to be preferred in presenting conclusions whenever it is possible that some further analysis may be made later, partly on the basis of the summary values being reported. Points (i), (iii), (v) and (vii) given above for graphical methods have fairly direct analogues for tabular presentation of conclusions and are there equally important.

4.7 Significance tests

We now turn to those types of analysis where the probability element is more central; the objective is to assess the uncertainty in the conclusions. Of course in some cases the important conclusions may be quite clear cut once the right approach to analysis is found. Thus with very extensive data, comparison of analyses of separate sections of data may be enough to establish the reliability of the general conclusions. We now, however, concentrate on the more formal assessment of uncertainty.

A widely used device for this is the significance test. We assume that the reader is familiar with the general argument involved: consistency with a null hypothesis H_0 is examined by calculating a P-value, the chance under the null hypothesis of obtaining a deviation as or more extreme than that observed.

There follow some miscellaneous comments on this procedure; these are best understood as experience of applications is gained.

(i) In practice it is rarely necessary to find P at all precisely. Often we can make a rough translation as follows:

$$P > 0.1, \text{ reasonable consistency with } H_0;$$
$$P \simeq 0.05, \text{ moderate evidence against } H_0;$$
$$P \leqslant 0.01, \text{ strong evidence against } H_0.$$

In reporting conclusions, the achieved P should, however, be given approximately. Note these remarks refer to a single test. The problem of combining information from several sets of data will be discussed later.

If $P \leqslant 0.05$, we say that the departure from H_0 is (statistically) significant at the 5 per cent level, etc.

(ii) A verbal definition of P is that it is the chance of getting a departure from H_0 as or more extreme than that observed, the chance being calculated assuming H_0 to be true. It may clarify this to give a more direct hypothetical physical interpretation to P. Suppose that we accepted the data under analysis as just convincing evidence against H_0. Then we would have to accept also more extreme sets of data as such evidence. Thus P is the long-run proportion of times in which we would falsely reject H_0 when it is in fact true, if we were to accept the data under analysis as just decisive against H_0. We stress that this is usually entirely hypothetical, one exception being the monitoring of routine testing.

(iii) The null hypothesis to be tested may arise in a number of ways. Sometimes it is a hypothesis thought quite conceivably to be true or very nearly so. For example, the null hypothesis might assert the absence of extra-sensory perception, the independence of certain genetic effects, the consistency of data with some physical theory, etc. A second kind of null hypothesis is considered not because it is in any way especially likely to be true, but because it divides the range of possibilities into two qualitatively different types. Thus a parameter θ may measure the difference between the mean yields of two alternative industrial processes. There may be no reason for thinking that $\theta = 0$, i.e. that the mean yields are the same, but the hypothesis that $\theta = 0$ may be of interest in that it divides the situations in which the first process has the higher mean yield from those in which it has the lower. So long as the data are reasonably consistent with H_0 the direction of the difference between the mean yields is not clearly established. In technological applications this second kind of null hypothesis is the more common.

One important use of significance tests is in the preliminary phase of an analysis as a guide to choice of models. It is useful there to distinguish between null hypotheses asserting simple primary structure, i.e. those that concern the primary aspects of the model, and null hypotheses asserting simple secondary structure, i.e. those that concern the secondary aspects of the model. As a simple example, suppose that it is hoped to interpret some data in the light of a linear regression relation between a response variable y and an explanatory variable x, particular interest focusing on the slope β of this relation. The null hypothesis that the regression is indeed linear is a hypothesis of simple primary structure: if evidence against the null hypothesis is found, the basis for the interpretation will be changed and different questions asked. By contrast, the null hypothesis of normality and constancy of variance would then be a hypothesis of simple secondary structure: evidence against the hypothesis would not change the objective of the analysis, although it might affect, conceivably to a major extent, the precise techniques of analysis to be used.

(iv) In order to develop a test, some knowledge of the kind of departure from H_0 of interest is essential. All sets of data will be exceptional in some respects. If several different kinds of departure are of interest, several tests applied and the most significant taken, it is essential to make an allowance for selection; see Section 3.7.

(v) A significance test is not directly concerned with the magnitude of any departure from H_0. It measures only consistency with the null hypothesis. The data may be consistent with the null hypothesis and at the same time be consistent with situations radically different in their practical implications. Thus in comparing the mean yields of two processes with limited data, the data might be consistent with zero difference, a large positive difference or a large negative difference between the two processes. On the other hand, sometimes a highly significant departure from H_0 may be of such small magnitude as to be for most purposes of little practical importance. For example, evidence against the inverse square law of gravitation was highly significant from data available from Newton's time onwards, i.e. there was strong evidence of some departure from the law. Yet the magnitude of the departure was so small that for most purposes the law was and is entirely adequate.

(vi) It should be clear from the above discussion that a significance test is not a procedure for deciding, say, which of two industrial processes should be recommended for use. This must depend on what other evidence is available and on economic considerations. All the significance test can do is to assess whether the data under analysis provide reasonably decisive evidence concerning the direction of any difference.

(vii) To summarize, significance tests have a valuable but limited role to play in the analysis of data. Rarely is a significance test the only analysis required. Nearly always we need also some idea of the magnitude of any departure from the null hypothesis that may be present.

The most widespread use of significance tests is in fields where random variation is substantial and where there is appreciable danger of premature claims that effects and relationships have been established from limited data. Significance tests play a valuable part in limiting such claims. Nevertheless, it is important that this rather negative role does not inhibit general judgement and initiative. This is one reason why estimation of effects is usually desirable as well as significance testing. Significance tests are concerned with questions like 'is such and such an effect reasonably firmly established by these data', and not with the question 'might there be something here well worth further study'.

4.8 Interval estimation

Suppose now that we are interested in some component parameter θ in a model. In some cases it will be enough to obtain a good point estimate of θ without considering explicitly the precision of the estimate. Now, however, we consider what is to be done when explicit calculation of precision is required.

A simple example related to the one above concerns data $(y_1, \ldots, y_n)$ represented by a model according to which the random variables $Y_1, \ldots, Y_n$

are independently normally distributed with unknown mean μ and known variance σ_0^2. A point estimate, such as the sample mean $\bar{y} = \Sigma y_i/n$, gives a value likely to be close to the unknown parameter μ. If we want to show the precision of the estimate it is natural to try to calculate intervals likely to contain the unknown μ.

Quite generally, with an unknown parameter θ in a specified model, a quantity $t^\alpha = t^\alpha(y)$ is called an upper $1-\alpha$ confidence limit for θ if, whatever the true value of θ,

$$\text{pr}\{\theta \leqslant T^\alpha = t^\alpha(Y)\} = 1-\alpha. \tag{4.9}$$

That is, the confidence limit is calculated by a procedure giving an upper limit that would be wrong only in a proportion α of trials in hypothetical repetitions.

We can define lower confidence limits in a similar way. Very often it is convenient to specify the uncertainty in the parameter by giving for a few values of α a $1-\frac{1}{2}\alpha$ upper limit and a $1-\frac{1}{2}\alpha$ lower limit, thereby forming a so-called $1-\alpha$ equi-tailed confidence interval. This is an interval of values calculated from the data in such a way that only in a proportion $\frac{1}{2}\alpha$ of hypothetical repetitions will it lie below the calculated lower limit, and similarly for the upper limit.

In the special case of the normal distribution mentioned above, the $1-\alpha$ confidence interval using the mean of the data is

$$\left(\bar{y}-k_{\frac{1}{2}\alpha}^*\frac{\sigma_0}{\sqrt{n}}, \bar{y}+k_{\frac{1}{2}\alpha}^*\frac{\sigma_0}{\sqrt{n}}\right), \tag{4.10}$$

where $\Phi(-k_{\frac{1}{2}\alpha}^*) = \frac{1}{2}\alpha$, $\Phi(.)$ denoting the standardized normal integral.

No attempt will be made here to cover the full theory of confidence-interval estimation. The following general points need watching in working with the various special cases which we shall encounter later.

(i) There is a close formal relation between confidence intervals and significance tests. We can look on a $1-\alpha$ confidence interval as the collection of possible parameter values that would not be judged inconsistent with the data at level α in a two-sided significance test.

(ii) Very often the confidence intervals are at least approximately symmetrical about some estimate t and the limits at various levels correspond to a normal distribution of standard deviation σ_t. Then it is convenient to say that θ is estimated by t with a standard error σ_t. This is to be regarded as a simple, concise way of describing a collection of confidence intervals at various levels. Occasionally, other concise descriptions are useful. Note that transformation of the parameter may help in this description by making the confidence limits more nearly symmetrical and more nearly described by Equation (4.10).

(iii) Very occasionally it is misleading to give an interval of values because, for example, two disjoint intervals of values may well be consistent with the data, the intermediate values being inconsistent. Then a generalized idea, namely of confidence regions, is called for.

(iv) If there are several ways of calculating intervals with the right probability of covering the true value, theoretical arguments are normally used to find which method produces the most sensitive analysis, in the sense of confidence intervals that are as selective as possible. Of course other considerations such as ease of computation and insensitivity to secondary assumptions in the model also enter.

(v) When, as is usually the case, there are nuisance parameters present, it is important that the probability that the random interval covers the true value should be at the required level, at least approximately, whatever may be the true value of the nuisance parameter.

(vi) The physical meaning of the confidence interval is given by the properties of the interval in hypothetical repetitions. The distinction between a confidence interval and a probability statement is fairly subtle and for some purposes unimportant.

(vii) Occasionally we work with intervals for future observations from the same random system. Prediction intervals can be defined in virtually the same way as for confidence intervals.

4.9 Decision procedures

The idea of statistical analysis as a tool for decision making in the face of uncertainty is important. It is nearly always fruitful to consider:

(i) What is the precise objective of the analysis?
(ii) What are the possible courses of action open?
(iii) What are the consequences of taking the 'wrong' decision?
(iv) What relevant evidence is available in addition to the data under analysis?

Where these aspects can be precisely formulated, and (iii) and (iv) specified quantitatively, a formula specifying the optimum decision corresponding to the data can be obtained. While there are important applications of quantitative decision theory (for example in control theory and industrial acceptance sampling), most statistical analyses are concerned rather with assessing the information in data and describing the conclusions which it is reasonable to draw. For this reason we shall not put much emphasis on procedures explicitly for decision making, although the broad points (i)–(iv) above are of wide qualitative importance, particularly in giving sharp focus to the analysis; see Examples D and S concerned with proportions of observations outside tolerance limits.

4.10   Examination of the adequacy of models

The conclusions drawn from a probabilistically based analysis will typically be phrased in terms of the parameters of one (or more) models, these models being provisionally treated as correct. Checking on the adequacy of the model is important not only in the definitive stage of the analysis, but more particularly in the preliminary phase when we are aiming to develop an appropriate model. There are broadly five ways of examining the adequacy of a model, although some of these ways are quite closely related. They are:

(i) the calculation of discrepancies between observed values of key responses and values fitted from the model. These discrepancies, usually called residuals, can be calculated so that if the model is adequate, the residuals should be very nearly completely random. As in the regression problem of Example G, plots of the residuals looking for systematic structure give a graphical test of the model;

(ii) we may proceed as in (i), but inspect a table of observed and fitted values and residuals to see whether the nature of the discrepancies gives a clue as to model inadequacy. Observed and fitted frequencies are compared in Examples W and X;

(iii) an overall test statistic may be calculated measuring the discrepancy between observed and fitted values and, assuming that its distribution, under the null hypothesis that the model is adequate, can be found, a test of significance results. See, for instance, Example P;

(iv) the model may, as in Example T, be expanded by the insertion of one or more additional parameters representing departures thought of potential interest, these extra parameters estimated and their magnitude and the statistical significance of the departures assessed. This is essentially equivalent to examining particular aspects of the residual configuration;

(v) a quite different kind of model can be fitted and the adequacy of fit of the two models compared in some way.

Method (iv) is to be preferred when there is some fairly clear notion of important types of departure from the model likely to arise, whereas methods (i)–(iii) are of more use for suggesting previously unanticipated features.

4.11   Parameters and parameterization

We now comment on the role of parameters in probability models and probabilistic statistical analysis. Some of these points belong to Chapter 3 on strategical issues, but they have been postponed to here because of their rather more technical character.

The physical interpretation of a parameter is sometimes as giving a property of a well-defined population of individuals, but more commonly as giving a property of a hypothetical system obtained by repetition under similar conditions. As such, parameters, at least those of primary interest, are intended to represent in idealized form important properties of the system under investigation. Sometimes, however, as explained in Section 2.3 (ii), the parameters are of main interest as an intermediate step in the prediction of future observations from the same or related random system.

Use of parameters in this way seems essential if relatively complex data are to be described concisely, if comparisons are to be made conveniently between a considerable number of sets of similar data and if sensitive probabilistically based comparisons are to be made.

Suppose, for example, that we have a fairly large set of data on a response variable y measured, say, for two different treatments and that no explanatory variables, other than treatment, are available. It will then be sensible to estimate, e.g. by a histogram, the frequency distribution of the y in each treatment group and/or to give the corresponding cumulative frequency curves. This might well be the simplest and best way of presenting the data. Confidence bands and other indicators of precision could if required be calculated. But if:

(i) there are many such sets of data, comparing the same two treatments under rather different conditions; or

(ii) there are other explanatory variables to be considered simultaneously; or

(iii) great interest attaches to the comparison of the location of the two distributions and to assessing the precision with which this comparison can be made;

then it will normally be good to introduce a parametric formulation. Thus in (i) there would be at least one parameter for each set of data defining the difference between treatments; some interest would probably attach to the null hypothesis that these parameters are the same for all sets of data, and if evidence against this is obtained there may be need for a second stage of modelling in attempting to explain how the differences vary between data sets. In (iii) again we may introduce a parameter for the difference in location, and perhaps other parameters for other aspects of the comparison, and consider the efficient estimation with confidence limits of that difference parameter; see Example F.

This is broadly the reason why parametric representation of the primary aspects of the problem will often be desirable. Parametric representation of the secondary aspects of the problem is a rather different issue.

In the above example, if previous experience and inspection of the data suggest it to be reasonable, it would often be very convenient and sensible to make the secondary assumption that the random variation is normally distributed, perhaps even with constant variance. Of course, in particular contexts special parametric distributions other than the normal might be called for;

an exponential distribution is taken in Example U. It would be important that the assumption say of normality, was not strongly contradicted by the data (Example P) and that the conclusions did not depend critically on the assumption of normality; in fact, provided due care is taken in checking for the influence of isolated extreme observations, the assumption of normality, possibly after transformation of the responses, will quite often in practice be an extremely convenient simplifying fiction. Of course, if attention is focused on the tails of the distribution, e.g. on the chance of long survival in a medical context, the assumption of a parametric distributional form may have to be looked at very critically.

If it has been decided to represent some features of a system by unknown parameters in a parametric model, there remains the choice of the particular mathematical form for the parameters concerned. As a simple example, the equation of simple linear regression,

$$E(Y_i) = \beta_0 + \beta_1 x_i, \tag{4.11}$$

can be written in the alternative forms

$$E(Y_i) = \gamma_0 + \gamma_1(x_i - \bar{x}), \tag{4.12}$$

and

$$E(Y_i) = \delta_1(x_i - \delta_0); \tag{4.13}$$

here $\bar{x} = \Sigma x_i / n$, so that whereas β_0 is the value of $E(Y)$ at $x = 0$, γ_0 is the value of $E(Y)$ at $x = \bar{x}$, the data mean, and δ_0 is the value of x for which $E(Y) = 0$. Now the relations (4.11)–(4.13) are exactly equivalent; with the proviso that the slope is nonzero, we can pass from any one form to any other by an appropriate transformation of parameter values, e.g.

$$\gamma_1 = \beta_1, \ \gamma_0 = \beta_0 + \gamma_1 \bar{x}; \ \delta_1 = \beta_1, \ \delta_0 = -\beta_0/\beta_i. \tag{4.14}$$

Provided that the parameters are allowed to take arbitrary real values, the models are mathematically equivalent. We call the passage from one form to another reparameterization: infinitely many such changes of form are always possible.

Note that discussion of choice of parameterization presupposes that we have chosen a model. How is a parameterization to be chosen? There are a number of considerations, sometimes conflicting: when there is a serious conflict it may be necessary to use, for example, different parameterizations in carrying out the analysis and in presenting the conclusions.

There are broadly four things to aim for:

(i) physical interpretability of the individual component parameters;
(ii) stability of the parameter values over different similar systems;
(iii) simplicity of discussions of errors of estimation;
(iv) numerical-analytical stability of the estimation procedure.

The two prime considerations for the final interpretation are (i) and (ii). Often the slope is of main concern and then there is nothing to choose between Equations (4.11) and (4.12). But specific subject-matter considerations may indicate, for instance, that δ_0 is of special interest and that Equation (4.13) is the appropriate form. Alternatively, analysis of a large number of sets of data might show that, although they had different slopes and values of γ_0, all had the same intercept β_0 at $x = 0$ and that therefore Equation (4.11) is a valuable form.

Consideration (iii), the assessment of precision, points toward Equations (4.11) and (4.12), and especially to (4.12), where errors of estimation in γ_0 and γ_1 are independent. While numerical stability is not often critical in calculations as simple as that for fitting a straight line, problems would be encountered if model (4.11) were fitted directly when the origin is remote from the region of values of x covered by the data. The model in the form (4.12) would raise no such problems. The general moral is that both for obtaining estimates with errors having simple properties, and for reasons of numerical stability, fitting of models with approximately orthogonalized parameters is advantageous. The details of fitting are thus best carried out in that form, but, as stressed above, the more basic points of interpretation (i) and (ii) may dictate transformation of the results to a reparameterized form. Examples F and I both illustrate the advantages of careful parameterization in handling linear models.

There are further general points about model choice and parameter formulation that apply when a number of similar sets of data are under analysis. As stressed in (ii) above, it will be an advantage to find a parameterization in which as many component parameters as possible are effectively constant over data sets. Consider now a parameter of interest for which this is not possible: suppose, to be specific, that with a number of straight-line relationships the slope varies between data sets in an important way. The first objective, which may not be achievable, is to find a representation in which the variation between sets of data has been 'removed' or 'explained'. Possibilities include the following:

(a) it may be that transformation, e.g. to consideration of $\log y$ versus $\log x$, will lead to effectively linear relations of constant slope;

(b) it is possible that, if the ranges of the explanatory variable are different in different data sets, that the different straight lines are really all part of a single nonlinear function;

(c) it may be that there is a further explanatory variable z associated with each whole data set, that will 'explain' the variation in slope, e.g. by a representation

$$\beta_j = \phi + \psi(z_j - \bar{z}),$$

where β_j is the slope in the jth data set;

(d) it may be that the slopes can be divided into two or more sets, in a rational way, such that slope is constant in each set;

(e) a final possibility if (a)–(d) seem inapplicable is to suppose that the variation between sets of data is random; one might suppose that the true slopes vary between data sets in a normal distribution with unknown mean and variance.

All these possibilities are in effect devices for avoiding models with large numbers of parameters. Of these (e) is conceptually the least satisfactory, because it provides merely a description of the fact that slope varies in an unpredictable way; the other approaches (a)–(d) explain or predict the slope to be achieved in specified circumstances. We repeat that these considerations are relevant only when the variation in slope is judged important.

4.12 Transformations

Although in principle each major new statistical analysis deserves formulation from first principles, there are obvious arguments of computational economy and standardization of presentation in using wherever reasonably possible a standard model for which the techniques of analysis are readily available and well understood. In particular the normal-theory linear model, Example 2.3, has many attractions. Often, moreover, the analysis of complex sets of data can be pieced together by combining a number of relatively standard analyses in sections.

One powerful method for extending the range of applicability of standard methods is transformation, i.e. the taking of logs, powers and other nonlinear functions. This can be done in three distinct ways, namely by:

(i) a nonlinear transformation of an effectively continuous response variable to a new form;

(ii) a nonlinear transformation of the parameter, usually an expected value, in a representation of systematic variation;

(iii) a nonlinear transformation of an explanatory variable.

In (i) we might assume that some transformed form of the response variable satisfies a normal-theory linear model; see Example J, where taking logs is crucial, and Example G, where taking logs seems desirable. In determining a transformation we aim at achieving a simple linear structure, at achieving constant variance and at achieving normal distributions. In case of conflict between these aims, attention should be concentrated on the primary aspect of the model, usually, but not necessarily, the simplicity of the systematic structure. Choice of transformation is very often based on previous experience and preliminary analysis, but more systematic procedures are available by regarding a transformation parameter as one to be estimated.

In (ii) the nonlinear transformation is applied not to the response but to a parameter, usually the mean, of its distribution. Thus in analysing binary data, the expected response is the probability of 'success', and an important family of models is obtained by supposing that

$$\log \left\{ \frac{\text{pr (success)}}{\text{pr (failure)}} \right\}$$

obeys a linear model; see Examples V, W and X.

The distinction between (i) and (ii) is illustrated in the context of normal theory by supposing that:

(a) some function $g(Y_i)$ of the response Y_i, e.g. $g(Y_i) = \log Y_i$, is normally distributed with constant variance, and with mean having some simple structure in terms of the explanatory variables;

(b) Y_i is normally distributed with constant variance and that some function $g\{E(Y_i)\}$ of its mean, e.g. $\log E(Y_i)$, has simple structure in terms of the explanatory variables.

In this particular context, (a) is the procedure almost always used.

Linear transformations of response and explanatory variables, i.e. the use of a new origin and unit of measurement, will usually bring about no essential change in the model and its interpretation, but such changes may often be desirable for numerical analytical reasons, to avoid rounding errors. With a considerable number of explanatory variables it will be a wise precaution to scale them to have in the data approximately zero mean and unit standard deviation. This is for computation; for interpretation one would normally return to the original units.

4.13 Interaction

Finally we discuss some qualitative aspects of the important concept of interaction and more specifically the notion of absence of interaction. These aspects concern the systematic relation between a response variable and two or more explanatory variables and are best discussed by considering idealized situations in which random variation is absent.

Suppose initially that there are two explanatory variables, x_1 and x_2. The discussion and interpretation are rather different depending on whether these are:

(i) variables representing treatments; or
(ii) variables representing intrinsic properties of the individuals;
and
(a) take a continuous range of variables; or
(b) take a number of qualitatively different values.

Absence of interaction between x_1 and x_2 means that the true response $\eta(x_1, x_2)$ is such that the difference of η between any two levels of x_2 is the same

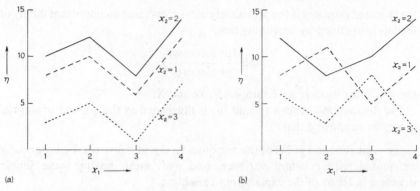

Fig. 4.1. Response to two factors x_1, x_2. (a) No interaction; (b) interaction

for all levels of x_1. If, say, x_1 is continuous this means that if η is plotted against x_1 for fixed x_2, the resulting curves are parallel for different levels of x_2; see Fig. 4.1. If both x_1 and x_2 are quantitative, the curves are parallel also when plotted against x_2 for fixed x_1. Note that if one variable is a treatment and the other an intrinsic property, we would normally plot against the former for various levels of the latter. Mathematically the requirement is that $\eta(x_1, x_2) = \eta_1(x_1) + \eta_2(x_2)$. If the levels are qualitative, absence of interaction is exemplified by the artificial data of Table 4.3, in which the difference between two rows is the same for all columns and the difference between two columns the same for all rows.

Interaction is any departure from this condition: of course with random variation no set of data can be expected to accord exactly with the structure of Table 4.3 and we are normally interested mainly in departures from the simple no-interaction structure too large to be reasonably accounted for by chance.

There are two rather different reasons why absence of interaction is important. First, if both x_1 and x_2 represent treatments, the discussion cf their effects is simplified; in fact in Table 4.3 the marginal means specify the changes in response as we change the levels of one variable holding the other fixed, and if

Table 4.3. Fictitious example of absence of interaction for two qualitative variables. Values are responses η in absence of random variation

Level of x_2	Level of x_1				
	1	2	3	4	Mean
1	8	10	6	12	9
2	10	12	8	14	11
3	3	5	1	7	4
Mean	7	9	5	11	

we change both variables the separate effects are additive. There is sometimes also a reasonable implication that because the effects of the two treatment variables add, their effects are in some sense physically independent.

If, however, x_1, say, represents a treatment and x_2 an intrinsic property of the individuals, absence of interaction has a slightly different implication. The effect of changing levels of treatment does not depend on x_2 and this means that, for some purposes at least, x_2 can be disregarded. This both simplifies interpretation and strengthens the base for extrapolation to new individuals. In a sense, absence of appreciable interaction is a key element in rational extrapolation. Thus, suppose that there are two levels of x_1 representing two alternative medical or industrial treatments and that a limited number of levels of x_2 have been investigated. If there is no interaction, i.e. if the treatment difference is always the same, there is some hope that the same difference will apply to individuals with new values of x_2. On the other hand, if there is appreciable interaction, and especially if different treatments are preferable at different levels of x_2, extrapolation of any kind may be hard and in any case interpretation is considerably more complicated. Thus investigation for possible interactions in such contexts can be very important.

If interaction is present, it must be described, and if possible interpreted, in as simple a fashion as seems possible. Inspection of graphs like Fig. 4.1 or tables like Table 4.3 will normally be an essential first step. It would involve too much detail to develop fully ways of describing interaction. Among the more important possibilities are the following:

(i) to transform from η to some function such as $\sqrt{\eta}$, $\log \eta$, η^{-1}, etc., chosen so that on the transformed scale there is no interaction. Note, for example, that had we started in Table 4.3 with squares of the values given there, interaction would have been present. This interaction could then have been 're-moved' by taking square roots. We might then carry out the whole interpretation on the new scale, although in some circumstances, as when the original scale is extensive, transformation back to the original scale may be desirable at the end. Of course, only rather special patterns of interaction can be removed in this way;

(ii) when x_1 and x_2 are both quantitative, to build up a fairly simple mathematical form for the function $\eta(x_1, x_2)$;

(iii) especially if x_1 and x_2 take qualitatively different levels, to recognize individual cells (i.e. combinations of x_1 and x_2), or individual values of x_1 or of x_2 which depart from a general pattern of no interaction. An extreme example of the first possibility is that all combinations of x_1 and x_2, except one, give the same response. The situation is then most aptly described by specifying the unique combination and the two levels of response.

We have above sketched two rather different interpretations of interaction, the second indicating whether a treatment effect is independent of an intrinsic variable. There is a third possible use of interaction. Suppose that the levels of the factor x_2 correspond to an aspect which varies in an unstructured and largely haphazard way; for example, the levels may correspond to different sets of apparatus, different observers, different unidentified batches of experimental material, etc. Then interaction of treatment effect with x_2 that cannot be removed by transformation is in some sense random variation and the magnitude of such interaction may provide some basis for assessing the precision of the estimated treatment effect. In particular, in terms of the techniques of analysis of variance an interaction mean square may under such circumstances be used in the estimation of the error of treatment contrasts.

The previous discussion has been for two explanatory variables, where in fact very simple techniques of fitting and display are often adequate.

With more than two explanatory variables the notion of interaction becomes more complicated but in a sense more useful. With three explanatory variables, we consider two-factor interactions $x_2 \times x_3$, $x_3 \times x_1$, $x_1 \times x_2$, the last, for example, being defined by averaging response over a fixed set of levels of x_3 to form a table such as Table 4.3. Similarly a three-factor interaction can be formed that in effect measures how much the pattern of a two-factor interaction shifts from one level to another of the third factor. The highly developed technique of analysis of variance allows the relatively straightforward computation of sums of squares of deviations associated with interactions.

For interpretation, broadly the same points apply as for two factors. Absence of interaction between treatments allows simple description and interpretation via the marginal averages; presence of interaction always calls for more detailed interpretation. Absence of interaction between treatments and intrinsic variables means that treatment effects can be studied regardless of the value of the intrinsic variable. Sometimes interactions can be used to estimate the magnitude of the random variation present. Several of the examples in Part II concerned with factorial experiments illustrate these points. In particular, see Examples J and O for interpretation without interactions; Examples L, R and W illustrate the presence of interaction.

Part II   Examples

Example A

Admissions to intensive care unit

Description of data. Table A.1 gives arrival times of patients at an intensive care unit. The data were collected by Dr A. Barr, Oxford Regional Hospital Board. Interest lies in any systematic variations in arrival rate, especially any that might be relevant in planning future administration.

General considerations. These data represent point events occurring in time. A very simple model for comparison with such data is the completely random series or Poisson process. In this events occur (i.e. patients arrive) independently, the chance that an event occurs in a short period of length h being ρh, where ρ is a single parameter, the arrival rate. Occurrences in different time

Table A.1. Arrival times of patients at intensive care units. (To be read down the columns.)

1963			1963			1963			1963		
M	4 Feb.	11.00 hr	S	6 Apr.	22.05 hr	W	5 June	22.30 hr	T	23 July	21.45 hr
		17.00	T	9 Apr.	12.45	M	10 June	12.30	W	24 July	21.30
F	8 Feb.	23.15			19.30			13.15	S	27 July	0.45
M	11 Feb.	10.00	W	10 Apr.	18.45	W	12 June	17.30			2.30
S	16 Feb.	12.00	Th	11 Apr.	16.15	Th	13 June	11.20	M	29 July	15.30
M	18 Feb.	8.45	M	15 Apr.	16.00			17.30	Th	1 Aug.	21.00
		16.00	T	16 Apr.	20.30	Su	16 June	23.00	F	2 Aug.	8.45
W	20 Feb.	10.00	T	23 Apr.	23.40	T	18 June	10.55	S	3 Aug.	14.30
		15.30	Su	28 Apr.	20.20			13.30			17.00
Th	21 Feb.	20.20	M	29 Apr.	18.45	F	21 June	11.00	W	7 Aug.	3.30
M	25 Feb.	4.00	S	4 May	16.30			18.30			15.45
		12.00	M	6 May	22.00	S	22 June	11.05			17.30
Th	28 Feb.	2.20	T	7 May	8.45	M	24 June	4.00	Su	11 Aug.	14.00
F	1 Mar.	12.00	S	11 May	19.15			7.30	T	13 Aug.	2.00
Su	3 Mar.	5.30	M	13 May	15.30	Tu	25 June	20.00			11.30
Th	7 Mar.	7.30	T	14 May	12.00			21.30			17.30
		12.00			18.15	W	26 June	6.30	M	19 Aug.	17.10
S	9 Mar.	16.00	Th	16 May	14.00	Th	27 June	17.30	W	21 Aug.	21.20
F	15 Mar.	16.00	S	18 May	13.00	S	29 June	20.45	S	24 Aug.	3.00
S	16 Mar.	1.30	Su	19 May	23.00	Su	30 June	22.00	S	31 Aug.	13.30
Su	17 Mar.	11.05	M	20 May	19.15	T	2 July	20.15	M	2 Sept.	23.00
W	20 Mar.	16.00	W	22 May	22.00			21.00	Th	5 Sept.	20.10
F	22 Mar.	19.00	Th	23 May	10.15	M	8 July	17.30	S	7 Sept.	23.15
Su	24 Mar.	17.45			12.30	T	9 July	19.50	Su	8 Sept.	20.00
		20.20	F	24 May	18.15	W	10 July	2.00	T	10 Sept.	16.00
		21.00	S	25 May	21.05	F	12 July	1.45			18.30
Th	28 Mar.	12.00	T	28 May	21.00	S	13 July	3.40	W	11 Sept.	21.00
		12.00	Th	30 May	0.30			4.15	F	13 Sept.	21.10
S	30 Mar.	18.00	S	1 June	1.45			23.55	Su	15 Sept.	17.00
T	2 Apr.	22.00			12.20	S	20 July	3.15	M	16 Sept.	13.25
		22.00	M	3 June	14.45	Su	21 July	19.00	W	18 Sept.	15.05

(continued on p. 54)

1963			1963			1963–64			1964		
S	21 Sept.	14.10 hr	T	12 Nov.	7.45 hr	Su	15 Dec.	1.15 hr	S	25 Jan.	13.55 hr
M	23 Sept.	19.15	F	15 Nov.	15.20	M	16 Dec.	1.45	W	29 Jan.	21.00
T	24 Sept.	14.05			18.40	T	17 Dec.	18.00	Th	30 Jan.	7.45
		22.40			19.50	F	20 Dec.	14.15	F	31 Jan.	22.30
F	27 Sept.	9.30	S	16 Nov.	23.55			15.15	W	5 Feb.	16.40
S	28 Sept.	17.30	Su	17 Nov.	1.45	S	21 Dec.	16.15			23.10
T	1 Oct.	12.30	M	18 Nov.	10.50	Su	22 Dec.	10.20	Th	6 Feb.	19.15
W	2 Oct.	17.30	T	19 Nov.	7.50	M	23 Dec.	13.35	F	7 Feb.	11.00
Th	3 Oct.	14.30	F	22 Nov.	15.30			17.15	T	11 Feb.	0.15
		16.00	S	23 Nov.	18.00	T	24 Dec.	19.50			14.40
Su	6 Oct.	14.10			23.05			22.45	W	12 Feb.	15.45
T	8 Oct.	14.00	Su	24 Nov.	19.30	W	25 Dec.	7.25	M	17 Feb.	12.45
S	12 Oct.	15.30	T	26 Nov.	19.00			17.00	T	18.Feb.	17.00
Su	13 Oct.	4.30	W	27 Nov.	16.10	S	28 Dec.	12.30			18.00
S	19 Oct.	11.50	F	29 Nov.	10.00	T	31 Dec.	23.15			21.45
Su	20 Oct.	11.55	S	30 Nov.	2.30	Th	2 Jan.	10.30	W	19 Feb.	16.00
		15.20			22.00	F	3 Jan.	13.45	Th	20 Feb.	12.00
		15.40	Su	1 Dec.	21.50	Su	5 Jan.	2.30	Su	23 Feb.	2.30
T	22 Oct.	11.15	M	2 Dec.	19.10	M	6 Jan.	12.00	M	24 Feb.	12.55
W	23 Oct.	2.15	Tu	3 Dec.	11.45	T	7 Jan.	15.45	T	25 Feb.	20.20
S	26 Oct.	11.15			15.45			17.00	W	26 Feb.	10.30
W	30 Oct.	21.30			16.30			17.00	M	2 Mar.	15.50
Th	31 Oct.	3.00			18.30	F	10 Jan.	1.30	W	4 Mar.	17.30
F	1 Nov.	0.40	Th	5 Dec.	10.05			20.15	F	6 Mar.	20.00
		10.00			20.00	S	11 Jan.	12.30	T	10 Mar.	2.00
M	4 Nov.	9.45	S	7 Dec.	13.35	Su	12 Jan.	15.40	W	11 Mar.	1.45
		23.45			16.45	T	14 Jan.	3.30	W	18 Mar.	1.45
T	5 Nov.	10.00	Su	8 Dec.	2.15			18.35			2.05
W	6 Nov.	7.50	M	9 Dec.	20.30	W	15 Jan.	13.30			
Th	7 Nov.	13.30	W	11 Dec.	14.00	F	17 Jan.	16.40			
F	8 Nov.	12.30	Th	12 Dec.	21.15	Su	19 Jan.	18.00			
S	9 Nov.	13.45	F	13 Dec.	18.45	M	20 Jan.	20.00			
		19.30	S	14 Dec.	14.05	T	21 Jan.	11.15			
M	11 Nov.	0.15			14.15	F	24 Jan.	16.40			

periods are statistically independent. Such a model in a sense shows total lack of systematic structure. Among its properties are that the number of arrivals in a particular time length l has a Poisson distribution with variance equal to its mean, ρl; also the intervals between successive arrivals have an exponential distribution. More complex models allow for nonexponential distributions, dependence between successive intervals, etc.

Now in the present example it is unlikely that the Poisson process will be adequate; in particular, variations in rate with day of week and time of day are likely to be present and will be of interest in, for example, planning studies. The simple Poisson process is nevertheless valuable in judging the precision of comparisons, e.g. between different days of the week. Systematic variations in rate could be studied by formal explicit fitting of models in which ρ is not constant, but it is more direct to proceed by elementary descriptive methods. Thus we calculate separate rates for Sunday, Monday, ..., and similarly separate rates for each two-hourly period 00.00–, 2.00–, ... For the latter, when we pool over days of the week, there is an implicit assumption that any time-of-day pattern is similar for the different days. The Poisson distribution provides a standard against which the statistical significance of the observed variation can be judged.

The analysis. Variations of rate with time of day and day of week, and also longer-term variations, are examined in Tables A.2–A.4 by forming total numbers of occurrences in relevant groupings. Two-way tables, not given here, show no strong suggestion that, for example, any time-of-day variation is different on different days of the week.

Table A.2. Time-of-day variation

	No. of arrivals	Rate per 24 hr		No. of arrivals	Rate per 24 hr
0.00–	14	0.411	12.00–	31	0.910
2.00–	17	0.499	14.00–	30	0.880
4.00–	5	0.147	16.00–	36	1.056
6.00–	8	0.235	18.00–	29	0.851
8.00–	5	0.147	20.00–	31	0.910
10.00–	25	0.733	22.00–	23	0.675

Table A.3. Day-of-week variation

	No. of weeks	No. of arrivals	Fitted freq.	Rate per day
Mon.	59	37	36.64	0.627
Tue.	59	53	36.64	0.898
Wed.	59	35	36.64	0.593
Thu.	58	27	36.02	0.466
Fri.	58	30	36.02	0.517
Sat.	58	44	36.02	0.759
Sun.	58	28	36.02	0.483

$\chi_6^2 = 14.20$: omitting Tue., $\chi_5^2 = 6.12$.

Table A.4. Long-term variation

	Days	No. of arrivals	Fitted freq.	Rate per day		Days	No. of arrivals	Fitted freq.	Rate per day
Feb. 63	25	13	15.52	0.520	Sept. 63	30	17	18.63	0.567
Mar. 63	31	16	19.25	0.516	Oct. 63	31	17	19.25	0.548
Apr. 63	30	12	18.63	0.400	Nov. 63	30	28	18.63	0.933
May 63	31	18	19.25	0.581	Dec. 63	31	32	19.25	1.032
June 63	30	23	18.63	0.767	Jan. 64	31	23	19.25	0.742
July 63	31	16	19.25	0.516	Feb. 64	29	17	18.01	0.586
Aug. 63	31	15	19.25	0.484	Mar. 64	18	7	11.18	0.389

$\chi_{13}^2 = 21.82$.

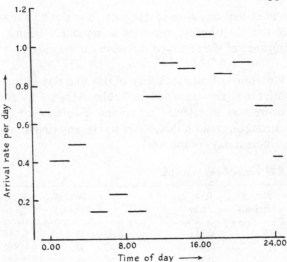

Fig. A.1. Patient arrival rate versus time of day.

Significance of variation above that to be expected from a Poisson distribution is tested by dispersion tests, i.e. in effect by comparison with fitted frequencies assuming constant rate. The time-of-day variation is overwhelmingly significant; the pattern of variation is best seen from Fig. A.1; the rate is highest between 12.00 and 22.00 hr and lowest between 4.00 and 10.00 hr. Variation between days of the week is just significant at 5 per cent; if Tuesdays, which have a high rate, are omitted, χ^2 drops to 6.12, with 5 degrees of freedom. Slower fluctuations, shown in Table A.4, are not quite significant at 5 per cent and show no very smooth pattern; the data do not really allow investigation of possible seasonal variation.

There are other features of the data that might be of interest in particular contexts. One such is the occurrence of short intervals between successive arrivals, which might be of special interest in an operational research study. Table A.5 gives the frequency of the shorter intervals. The mean interval is $(24 \times 409)/254 = 38.6$ hr and an exponential distribution with this mean gives

Table A.5. Short intervals between successive arrivals

	Observed freq.	Fitted freq. (exptl)	Fitted freq. (exptl × 1.25)
0 hr–	20	12.8	16.0
2 hr–	17	12.2	15.2
4 hr–6 hr	5	11.6	14.5

the frequencies in the second column. The exponential distribution corresponds, however, to a Poisson process of constant rate and the effect of variations in rate, such as those associated with time of day, is to increase the ordinate of the probability distribution at the origin by a factor approximately

$$1+c_\rho^2, \tag{A.1}$$

where c_ρ^2 is the coefficient of variation of rate. For the time-of-day variation $c_\rho^2 \simeq \frac{1}{2}$, and this leads to the modified frequencies shown. There is at least rough agreement with the approximate theory.

Further points and exercises

(i) Derive Equation (A.1). Calculate the whole frequency distribution of intervals between successive arrivals under a plausible model and compare with the empirical distribution.

(ii) Suppose that the data were grouped into 1 hr intervals and the number of arrivals in each such interval regarded as a discrete-time time series and a periodogram calculated. What form would it take? What would be the result if the time-of-day variation were sinusoidal but of different amplitude and phase on different days of the week?

(iii) Why would an analysis of the frequency distribution of intervals between successive arrivals not be adequate as the sole analysis of the data?

Related references. Armitage (1971, §7.7), Davies and Goldsmith (1972, §9.42), Snedecor and Cochran (1967, §§9.3, 9.4) and Wetherill (1967, §8.8) describe the dispersion test for Poisson variables. Cox and Lewis (1966) deal with the statistical analysis of data in the form of point events in time.

Example B

Intervals between adjacent births

Description of data. The data in Table B.1 were obtained by Greenberg and White (1963) from the records of a Genealogical Society in Salt Lake City. The distribution of intervals between successive births in a particular serial position is approximately log normal and hence geometric means are given. For example, the entry 39.9 in the bottom right-hand corner means that for families of 6 children, in which the fifth and sixth children are both girls, the geometric mean interval between the births of these two children is 39.9 months.

Table B.1. Mean intervals in months between adjacent births by family size and sequence of sex at specified adjacent births

Family size	Births	Sequence of sex			
		MM	MF	FM	FF
2	1–2	39.8	39.5	39.4	39.3
3	1–2	31.0	31.5	31.4	31.1
3	2–3	42.8	43.7	43.3	43.4
4	1–2	28.4	28.1	27.5	27.8
4	2–3	34.2	34.4	34.3	35.0
4	3–4	43.1	44.3	43.3	42.8
5	1–2	25.3	25.6	25.6	25.5
5	2–3	30.3	30.1	29.9	30.0
5	3–4	33.7	34.0	33.7	34.7
5	4–5	41.6	42.1	41.9	41.3
6	1–2	24.2	24.4	24.0	24.5
6	2–3	27.6	27.7	27.5	27.6
6	3–4	29.8	30.2	30.3	30.8
6	4–5	34.2	34.2	34.1	33.4
6	5–6	40.3	41.0	40.6	39.9

General considerations. This example illustrates a fairly common situation in which the data available for analysis are a substantial reduction of the originating data. One would really like access to further information.

The variation involved here is natural variation. The full data would consist at one level of a set of frequency distributions, one distribution for each combination of sequence number and sex pairing. In more detail one would have for each family a set of intervals and sexes. All that can be done with the available data is to describe as concisely as possible any systematic structure. This comes down in essence to studying and describing three things:

(i) systematic differences between the sex pairings MM, MF, FM, FF;
(ii) the variation with sequence number;
(iii) any simple patterns of interaction between (i) and (ii).

While these comparisons can and should be done at a descriptive level, i.e. by simple plotting and tabulation, it is certainly worth while also attaching an approximate measure of precision to important derived quantities.

Of course if the further data mentioned above were available, many possibilities for further analysis and interpretation become available. A comparison solely in terms of the values presented here will be complete only if

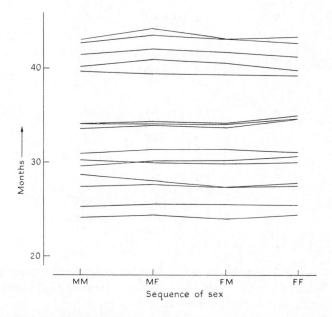

Fig. B.1. Mean interval versus sequence of sex. Each line represents one row of Table B.1, i.e. one family size, birth sequence combination.

the other features of the data, e.g. the shapes and dispersions of the distributions, are constant.

The analysis. There are three factors of interest: family size, birth order and sequence of sex. The sequence of sex (MM, MF, FM, FF) across each row of Table B.1 shows little variation and a simple plot (Fig. B.1) confirms the absence of any major such effect. Hence we can average over this factor. Also the four observations across each line of Table B.1 can be used to provide some indication of variability.

Since the distribution of intervals between births is approximately log normal, we calculate the average of log (months), although in the present circumstances taking logarithms is something of a refinement; throughout natural logs are used. The averages for each family size and birth interval are given in Table B.2. In all cases, the final interval is approximately $e^{3.7} \simeq 40$ months, earlier intervals being decreasingly shorter. The data of Table B.2 are

Table B.2. Average values of log months and corresponding antilogs

Family size	Births	Log months	Months
2	1–2	3.676	39.5
3	1–2	3.442	31.2
	2–3	3.768	43.3
4	1–2	3.330	27.9
	2–3	3.540	34.5
	3–4	3.770	43.4
5	1–2	3.239	25.5
	2–3	3.404	30.1
	3–4	3.527	34.0
	4–5	3.731	41.7
6	1–2	3.189	24.3
	2–3	3.318	27.6
	3–4	3.410	30.3
	4–5	3.526	34.0
	5–6	3.700	40.4

plotted in Fig. B.2, aligned according to final interval, and show how the intervals 2–3, 3–4, . . . lie roughly on a common curve, with the interval 1–2 for each family size displaced below it.

A sufficiently accurate estimate of variation is provided by the range across MM, MF, FM, FF. Across the first row of Table B.1, the range is log(39.8) −

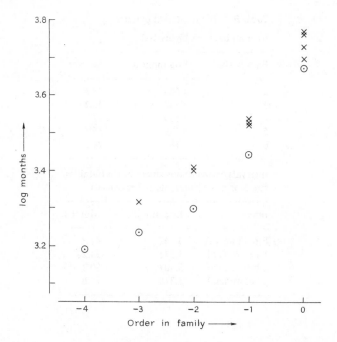

Fig. B.2. Mean of log birth interval versus order in family: 0, final interval; -1, previous interval; . . .

⊙   intervals (1–2)
✕   intervals (2–3), . . .

$\log(39.3) = 0.0126$. The average range over all 15 rows of data is 0.0216, giving an estimated standard deviation (Pearson and Hartley, 1972, Table 9) $\tilde{\sigma} = 0.0216 \times 0.4857 = 0.0105$; the range does not vary systematically.

The greatest scatter about the trend in Fig. B.2 lies in the final interval. Excluding that for family size 2, since the interval 1–2 lies below the common curve, the values of log(months) for the final interval are 3.768, 3.770, 3.731, 3.700, having a mean square of 0.001 11 on 3 degrees of freedom. This is much greater than $\tilde{\sigma}^2/4 = 0.000\ 028$.

There is little point in fitting a mathematical model to represent the trend. Instead we summarize the results by averaging the appropriate observed values; antilogs of these averages are given in Table B.3.

To attach an approximate standard error to the results is difficult. We notice from Table B.1 that the final interval for family sizes 5 and 6 is systematically shorter than 42.2 months, the result quoted Table in B.3, and for family sizes 3 and 4 it is correspondingly longer but the difference is small compared with variation in the overall pattern. None of the observed values differs from the pattern in Table B.3 by more than about 5 per cent.

Table B.3. Birth interval pattern

Interval between births 1–2

Family size	Log months	Months
2	3.676	39.5
3	3.442	31.2
4	3.330	27.9
5	3.239	25.5
6	3.189	24.3

Intervals between subsequent births (families size 3 or more, intervals 1–2 excluded)

Interval	Log months	Months
Final interval	3.742	42.2
1 before final	3.531	34.2
2 before final	3.407	30.2
3 before final	3.318	27.6

Example C

Statistical aspects of literary style

Description of data. As part of an investigation of the authorship of works attributed to St Paul, Morton (1965) found the numbers of sentences having zero, one, two, . . . occurrences of 'kai' (≡ and) in some of the Pauline works. Table C.1 gives a slightly condensed form of the resulting frequency distributions.

Table C.1. Frequency of occurrences of 'kai' in 10 Pauline works

Number of sentences with	Romans (1–15)	1st Corinth.	2nd Corinth.	Galat.	Philip.
0 kai	386	424	192	128	42
1 kai	141	152	86	48	29
2 kai's	34	35	28	5	19
3 or more kai's	17	16	13	6	12
No. of sentences	578	627	319	187	102
Total number of kai's	282	281	185	82	107

Number of sentences with	Colos.	1st Thessal.	1st Timothy	2nd Timothy	Hebrews
0 kai	23	34	49	45	155
1 kai	32	23	38	28	94
2 kai's	17	8	9	11	37
3 or more kai's	9	16	10	4	24
No. of sentences	81	81	106	88	310
Total number of kai's	99	99	91	68	253

General considerations. The raw data for each work form a frequency distribution, and in fact in the source paper a number of such distributions are given corresponding to various properties. Discussion is much simplified by replacing each distribution by a single number, a derived response variable. This might be obtained via a theoretical model for the distribution. In the present case, however, the mean seems the obvious candidate. We could use the ordinary mean number of kai's per sentence, e.g. 282/578 = 0.4879 for work I, or a modified mean in which 3 or more kais count as 3, 0.4498 for work I. In fact it makes little difference, although the second, which is used in the following analysis, has the advantage of being insensitive to the occurrence of a small number of sentences each with a large number of kais. The modified mean can be assigned a notional standard error in the usual way.

It is natural to see whether the works can reasonably be put into a small number of groups, all the works in the same group having effectively the same mean. To a large extent this is satisfactorily done in an informal way. A more elaborate procedure is to examine, at least in principle, all possible groupings of 2 works, of 3 works, . . . , testing the means in each possible grouping for consistency. The set of all groupings not 'rejected' at, say, the 5 per cent level constitutes a 95 per cent confidence set of possible groupings. Note especially that even though a large number of tests is involved, no adjustment of the significance level is required; the argument is simply that any 'correct' grouping will appear in the final list, except for the specified probability.

Any conclusion that the different groups were written by different authors involves a major further assumption.

To check whether useful information has been sacrificed by considering only the mean, it is worth examining another aspect of the distribution, and the standard deviation is the most amenable. To help interpret the relation between standard deviation and mean, it is a good idea to consider a very idealized stochastic model in which, within a work, each word has a constant probability, independently from word to word, of being 'kai'.

Finally, note that the consideration of any probabilistic aspects in this problem raises conceptual issues. The data available are not samples but complete enumerations and the works are unique, so that direct consideration of 'repetitions' of the data is not possible. The use of probabilistic ideas, like standard errors, depends on the tentative working hypothesis that for certain purposes the number of kai's per sentence behaves *as if* it were generated by a probabilistic mechanism. Put alternatively, we use the variation that would occur in random sampling as a reference standard to judge the observed variation. It is especially important in this problem that the final conclusion should make sense from a descriptive viewpoint and should not rely too heavily on the probabilistic analysis.

Some methodological details. If we consider k works with means $\bar{Y}_1, \ldots, \bar{Y}_k$ with standard errors $\sqrt{v_1}, \ldots, \sqrt{v_k}$, then the consistency of the k means is tested by the χ^2 statistic with $k-1$ degrees of freedom

Example C 65

$$\chi^2 = \Sigma\, (\bar{Y}_j - \bar{Y}.)^2/v_j$$
$$= \Sigma\, \bar{Y}_j^2/v_j - (\Sigma\, \bar{Y}_j/v_j)^2(\Sigma\, 1/v_j)^{-1}, \qquad \text{(C.1)}$$

where

$$\bar{Y}. = (\Sigma\, \bar{Y}_j/v_j)(\Sigma\, 1/v_j)^{-1}.$$

Because of the quite large numbers of observations involved we ignore errors of estimation of the v_j.

To help understand the values taken by the standard deviation of the distribution, suppose that sentences contain m words, and that the probability of any word being kai is θ, independently for different words. Then Y, the number of kai's in a sentence, has a binomial distribution with

$$E(Y) = m\theta, \ \text{var}(Y) = m\theta(1-\theta). \qquad \text{(C.2)}$$

Now suppose that sentence length is randomly distributed, i.e. that m is the value of a random variable M having mean and variance μ_m and σ_m^2. Then Equation (C.2) specifies the conditional mean and variance given $M = m$. It follows that unconditionally

$$E(Y) = \mu_m\theta, \ \text{var}(Y) = \mu_m\theta(1-\theta) + \sigma_m^2\theta^2. \qquad \text{(C.3)}$$

Even without direct information about the values of μ_m and σ_m^2 it follows on putting a range of plausible values into Equation (C.3) that the ratio of the variance to the mean is thus expected to be rather greater than 1 and is unlikely to be much less than 1. The use of modified means and standard deviations will tend to reduce the ratio a bit.

The analysis. Table C.2 gives for each work the mean, the modified mean, used in all the later analysis, the modified standard deviation, the ratio of variance

Table C.2. Summary statistics from 10 works

	Mean	Modif. mean $\bar{Y}$	Modif. st.dev. s	$s^2/\bar{Y}$	s.e.($\bar{Y}$)
I Rom.	0.4879	0.4498	0.7366	1.21	0.0306
II 1st Co.	0.4482	0.4306	0.7147	1.19	0.0285
III 2nd Co.	0.5799	0.5674	0.8171	1.18	0.0457
IV Galat.	0.4385	0.4064	0.6999	1.21	0.0512
V Phil.	1.0490	1.0098	1.0388	1.07	0.1029
VI Colos.	1.2222	1.1481	0.9632	0.81	0.1070
VII 1st Th.	1.2222	1.0741	1.1487	1.23	0.1276
VIII 1st Tim.	0.8585	0.8113	0.9473	1.11	0.0920
IX 2nd Tim.	0.7727	0.7045	0.8598	1.05	0.0917
X Heb.	0.8161	0.7742	0.9386	1.14	0.0533

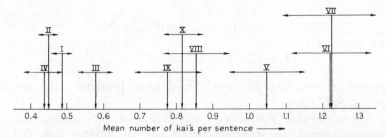

Fig. C.1. Mean number of kai's per sentence for 10 works. Arrows indicate plus and minus one standard error. I, Romans; II, 1st Corinth.; III, 2nd Corinth.; IV, Galat.; V, Philip.; VI, Colos.; VII, 1st Thessal.; VIII, 1st Timothy; IX, 2nd Timothy; X, Hebrews.

to mean, and the standard error of the mean. Figure C.1 shows the means and standard errors in convenient form.

The works fall into three groups, I–IV, V–VII and VIII–X, with some doubt about the consistency of III within the first group; otherwise the variations within groups are roughly in accord with the standard errors. The ratios of variance to mean are in agreement with the theoretical prediction, with the possible exception of work VI. Inspection of the frequency distribution for that work suggests that it would be unwise to put any strong interpretation

Table C.3. Some possible groupings of the works

Proposed grouping	Degrees of freedom	χ^2	Consistency at 5% level	Consistency at 1% level
I, II, III, IV	3	7.71	borderline	yes
I, II, IV	2	0.57	yes	yes
I, III, IV	2	6.51	no	yes
I, II, III	2	6.69	no	yes
II, III, IV	2	7.66	no	yes
I, III	1	4.56	no	yes
VIII, IX, X	2	0.72	yes	yes
III, VIII, IX, X	3	11.28	no	borderline
V, VIII, IX, X	3	5.49	yes	yes
VI, VIII, IX, X	3	11.73	no	no
VII, VIII, IX, X	3	5.94	yes	yes
V, VII, VIII, IX, X	4	9.69	borderline	yes
VI, VII, VIII, IX, X	4	15.46	no	no
V, VI, VIII, IX, X	4	14.82	no	no
V, VI, VII	2	0.87	yes	yes
V, VI, VII, VIII	3	6.43	no	yes
V, VI, VII, IX	3	11.88	no	no
V, VI, VII, X	3	13.94	no	no

Example C 67

on the anomalous dispersion. The rough agreement with a simple theory is some indirect confirmation of the use of probabilistic ideas.

To implement a more formal procedure, the χ^2 statistic (C.1) is found for some of the groups of potential interest. Table C.3 summarizes these calculations.

The simplest grouping consistent with the data is {I, II, IV}, {VIII, IX, X}, {V, VI, VII}, with III as an anomalous individual between the first two groups, if the calculations of precision are to be relied on. There are, however, alternative explanations, which, while not fitting so well, cannot definitely be excluded. Thus the anomalous work in the first group could be II or IV rather than III; alternatively, III could just conceivably be attached to the group {I, II, IV} or the the group {VIII, IX, X}. Or there are alternative possibilities for the last two groups; VI or VII or just possibly both could be attached to the second group, or VIII could be placed with {V, VI, VII}.

Further points and exercises

(i) Suggest a convenient parametric form for the distribution of M in Equation (C.3) and obtain the resulting distribution of Y. How would the data be analysed in the light of that distribution?

(ii) How would analysis be aided if μ_m and σ_m^2 were available for each work?

(ii) If one made the working assumption that an arrangement in three groups is to be obtained, how could the analysis be adapted so that the true combined split into three groups is included in the 'confidence set' with specified probability 95 or 99 per cent.

Example D

Temperature distribution in a chemical reactor

Description of data.* A chemical reactor has 1250 sections and it is possible to calculate a theoretical temperature for each section. These have a distribution across sections with mean 452 °C and standard deviation 22 °C; the distribution is closely approximated by a normal distribution in the range 390–520 °C. For a variety of reasons, measured temperatures depart from the theoretical ones, the discrepancies being partly random and partly systematic. The temperature is measured in 20 sections and Table D.1 gives the measurements and the corresponding theoretical values. It is known further that the measured temperature in a section deviates from the 'true' temperature by a completely random error of measurement of zero mean and standard deviation 3 °C. Special interest attaches to the number of channels in the reactor with 'true' temperature above 490 °C.

Table D.1. Measured and theoretical temperatures in 20 sections of reactor

Measured temp (°C)	Theoretical temp (°C)	Measured temp (°C)	Theoretical temp (°C)
431	432	472	498
450	470	465	451
431	442	421	409
453	439	452	462
481	502	451	491
449	445	430	416
441	455	458	481
476	464	446	421
460	458	466	470
483	511	476	477

General considerations. The theoretical temperatures t_{TH} are in principle known for the whole population and are the result of solving the appropriate partial differential equations for an idealized model of the reactor. They are

* Fictitious data based on a real investigation.

not random. The 'true' temperatures T depart from the theoretical temperatures for a variety of reasons, partly random, partly systematic. A simple working assumption is that over the whole reactor, treated as an effectively infinite population, we can treat the true and measured temperatures T and T_{ME} as random with

$$T = \alpha + \beta(t_{\mathrm{TH}} - t) + \varepsilon,$$

$$T_{\mathrm{ME}} = T + \varepsilon',$$

(D.1)

where t is the mean theoretical temperature in the measured channels, ε is a random term, normally distributed with zero mean and variance σ^2, and ε' is a measurement error of zero mean and variance 9, independent of T. The model has three unknown parameters α, β and σ^2.

Note that it is not assumed that the measured channels are a random sample from the population. It is, however, assumed that the linear model (D.1) holds over the whole range of t_{TH} and that the measured channels have ε's drawn from the appropriate distribution. There is no way of checking this latter point: to a limited extent linearity, constancy of variance and normality can be checked in the usual way.

Standard linear regression methods can be used to estimate the parameters and the variances of estimators. From the known distribution of t_{TH} it follows that over the reactor, treated as an infinite population, the distribution of T_{ME} is normal with mean and variance respectively

$$\xi' = \alpha + \beta(452 - \bar{t}), \qquad \eta = 484\beta^2 + \sigma^2 - 9$$

estimated by

$$\hat{\xi}' = \hat{\alpha} + \hat{\beta}(452 - \bar{t}), \qquad \hat{\eta} = 484\hat{\beta}^2 + \hat{\sigma}^2 - 9.$$

(D.2)

Note the subtraction of 9 because we are ultimately interested in T, not in T_{ME}.

The proportion of values of T above 490 °C is thus $\Phi(-\zeta)$, where

$$\zeta = \frac{490 - \xi'}{\sqrt{\eta}} = \frac{\xi}{\sqrt{\eta}}, \qquad \hat{\zeta} = \frac{\hat{\xi}}{\sqrt{\hat{\eta}}},$$

(D.3)

say. Rough confidence limits for the proportion, and hence for the number, are derived via a large-sample standard error for ζ, supplemented by a 'finite population adjustment'.

Methodological details. The parameters α, β and σ^2 are independently estimated by standard techniques and large-sample variances can be found: in particular, $\mathrm{var}(\hat{\sigma}^2) = 2\sigma^4/(n-1) \simeq 2\hat{\sigma}^4/(n-1)$, where $n = 20$ is the number of observations. Both $\hat{\xi}$ and $\hat{\eta}$ have negligible chance of being negative and it is thus convenient to consider

$$\text{var}(\log \hat{\xi}) \simeq \frac{\text{var}(\hat{\alpha}) + (452 - \bar{t})^2 \, \text{var}(\hat{\beta})}{\hat{\xi}^2},$$

$$\text{cov}(\log \hat{\xi}, \log \hat{\eta}) \simeq -\frac{2\hat{\beta} \times 484 \times (452 - \bar{t}) \, \text{var}(\hat{\beta})}{\hat{\xi}\hat{\eta}},$$

$$\text{var}(\log \hat{\eta}) \simeq \frac{4 \times 484^2 \hat{\beta}^2 \, \text{var}(\hat{\beta}) + \text{var}(\hat{\sigma}^2)}{\hat{\eta}^2},$$

leading to

$$\text{var}(\log \hat{\zeta}) = \text{var}(\log \hat{\xi} - \tfrac{1}{2} \log \hat{\eta})$$
$$\simeq \text{var}(\log \hat{\xi}) - \text{cov}(\log \hat{\xi}, \log \hat{\eta}) + \tfrac{1}{4} \text{var}(\log \hat{\eta}). \qquad \text{(D.4)}$$

From this, confidence limits for $\log \zeta$, and hence for ζ, and hence for $\Phi(-\zeta)$, follow. This gives approximate confidence limits for the proportion that would occur in a very large population: in 1250 channels, even if ζ were known and all other assumptions were satisfied, the number of affected channels would have a Poisson distribution of mean $\mu = 1250\Phi(-\zeta)$, so that a further error, one of prediction rather than estimation, has to be considered.

A relatively elaborate method for solving this prediction problem is outlined in Exercise (iii). A simple rough method uses the result that the variance of $\sqrt{N}$, where N has a Poisson distribution of mean μ, is $1/4$. Also,

$$\text{var}(\hat{\mu}) \simeq (1250)^2 \{\phi(-\hat{\zeta})\}^2 \, \text{var}(\hat{\zeta}),$$
$$\simeq (1250)^2 \{\phi(-\hat{\zeta})\}^2 \, \text{var}(\log \hat{\zeta})\hat{\zeta}^2$$

and

$$\text{var}(\sqrt{\hat{\mu}}) \simeq \tfrac{1}{4}\hat{\mu}^{-1} \, \text{var}(\hat{\mu}). \qquad \text{(D.5)}$$

Here $\phi(.)$ is the density of the standardized normal distribution. The effect of prediction error is thus to add $\tfrac{1}{4}$ to the right-hand side of Equation (D.5), leading to limits for $\sqrt{N}$ and hence for N.

The analysis. Standard linear regression of measured temperature on theoretical temperature leads to an estimated slope 0.5101, with estimated standard error 0.0865, to an estimated standard deviation about the regression line of 10.82 °C and to a sample mean measured temperature of 454.6 °C, with estimated standard error 2.42 °C. Given the population distribution of theoretical temperature, and the variance of measurement error, the distribution of true temperature is estimated to have mean

$$454.6 + 0.5101(452 - 459.7) = 450.67 \text{ °C}$$

and standard deviation

$$\sqrt{(484 \times 0.5101^2 + 10.82^2 - 3^2)} = 15.30 \text{ °C}.$$

Thus the standardized deviate for determining the proportion of channels above 490 °C in true temperature is

$$\hat{\zeta} = \frac{490 - 450.67}{15.30} = 2.571$$

and an estimate of the infinite population rate per 1250 channels is

$$1250\Phi(-2.571) = 6.3.$$

If a 97.5 per cent upper confidence limit is calculated via the large-sample standard error of log $\hat{\zeta}$, leading to normal confidence limits for log ζ, with derived limits for $\Phi(-\zeta)$, the upper limit is 28.3; if the calculation is done via $\sqrt{\Phi(-\zeta)}$, the limit is 23.5. In fact, the large-sample variance of $\sqrt{\hat{\mu}} = \sqrt{1250}\sqrt{\Phi(-\hat{\zeta})}$ is 1.42 and if this is inflated to allow for prediction error in the finite population this becomes 1.67, and the corresponding upper limit is raised from 23.5 to 25.5. The lower limits are less than 1.

Of course a very considerable extrapolation is involved in these calculations. The most it is reasonable to conclude is that, provided a reasonably stable linear relation holds over the whole temperature range, there are likely to be some channels with temperature above 490 °C, there are unlikely to be more than 30, and that a 'point estimate' of the number is 6.

Further points and exercises

(i) Outline how a full list of values of t_{TH} might be used to obtain a more refined estimate. Note that the 20 measured channels have negligible chance of their T being above 500 °C. For the remaining 1230 channels compute an estimated probability of the event of interest. How could precision be assessed?

(ii) Investigate the use of the noncentral Student t distribution to avoid the approximations involved in calculating confidence limits for $\Phi(-\zeta)$.

(iii) Develop a Bayesian approach to the calculation of prediction limits for the finite population problem via a gamma posterior distribution for the infinite population probability, leading to limits based on the negative binomial distribution.

Example E   A 'before and after' study
of blood pressure

Description of data. Table E.1 gives, for 15 patients with moderate essential hypertension, supine systolic and diastolic blood pressures immediately before and two hours after taking 25 mg of the drug captopril. The data were provided by Dr G. A. MacGregor, Charing Cross Hospital Medical School; for a report on the investigation and appreciable further summary data, see MacGregor, Markandu, Roulston and Jones (1979).

Table E.1. Blood pressures (mm Hg) before and after captopril

Patient no.	Systolic			Diastolic		
	before	after	difference	before	after	difference
1	210	201	−9	130	125	−5
2	169	165	−4	122	121	−1
3	187	166	−21	124	121	−3
4	160	157	−3	104	106	2
5	167	147	−20	112	101	−11
6	176	145	−31	101	85	−16
7	185	168	−17	121	98	−23
8	206	180	−26	124	105	−19
9	173	147	−26	115	103	−12
10	146	136	−10	102	98	−4
11	174	151	−23	98	90	−8
12	201	168	−33	119	98	−21
13	198	179	−19	106	110	4
14	148	129	−19	107	103	−4
15	154	131	−23	100	82	−18

General considerations. This example illustrates in skeleton form a number of points arising in the analysis of 'before and after' studies, i.e. investigations in which some property is measured on each individual before 'treatment', a treatment then applied and then the same property re-measured. The object is to assess the effect of treatment on the property in question.

It is common in such studies to have a control group receiving a placebo or dummy treatment, choice between treatment and placebo being randomized with due concealment. Where, as in the present instance, there is no control group, a comparison 'after' versus 'before' can still be made, but the interpretation of any difference as attaching specifically to the treatment is to be made with some reserve. When a control is omitted, it will be important to look for other information on the stability of the measured response; MacGregor *et al.* (1979) reported that mean blood pressure was stable both before and after treatment, and that these stable levels were different; this makes it less likely, although not impossible, that the difference reported below is a placebo effect.

Consider now the measurement of response, taking first, say, the systolic blood pressure. The simplest approach, and one commonly used, is to define as the derived response of interest the difference 'after' minus 'before' for each individual and then to analyse these differences, in particular comparing their mean with zero; it might be desirable to transform, e.g. logarithmically, first. In this the individual measurements before and after are discarded.

This is certainly legitimate: in the present instance the change in a patient's blood pressure is of immediate relevance both as a basis for analysing the full set of data and as an index of 'success' for that individual. Nevertheless, some information may be lost by concentrating on differences. First, if measurements before and after are almost independent, and we have encountered this in applications, there is the possibility either that substantial random variability of measurement of response is present or that ways of describing the data other than by differences would be preferable: for instance, if treatment reduced a very variable initial measurement to a practically fixed final level, an analysis solely in terms of differences would be seriously incomplete.

More generally, there is the possibility that the 'treatment' effect is different at different levels of initial measurement. This can be studied by plotting or regressing the difference 'after' minus 'before' versus the 'before' measurement. If, however, there are appreciable random variations in the measurements of response within an individual, caution is needed and the above simple procedure could be misleading.

To see this, suppose that for the ith individual the 'true' initial measurement is ξ_i, distributed across individuals with mean μ and variance σ_ξ^2, and that the initial measurement is

$$X_i = \xi_i + \varepsilon_i'; \tag{E.1}$$

here ε_i' is a random term of zero mean and variance σ_ε^2. Suppose that the effect of treatment is to add to the 'true' measurement

$$\Delta + \beta(\xi_i - \mu) + \eta_i, \tag{E.2}$$

where η_i is a random term of zero mean and variance σ_η^2 and that the final measurement Y_i is subject to an error ε_i'' having the same distribution as ε_i'.

Then

$$Y_i = \xi_i + \Delta + \beta(\xi_i - \mu) + \eta_i + \varepsilon_i'' \tag{E.3}$$

and the difference Z_i, 'after' minus 'before', is

$$Z_i = \Delta + \beta(\xi_i - \mu) + \eta_i + \varepsilon_i'' - \varepsilon_i'. \tag{E.4}$$

Assuming that all random terms are independent, it follows in particular that the regression coefficient of Z_i on X_i is

$$\frac{\operatorname{cov}(Z_i, X_i)}{\operatorname{var}(X_i)} = \frac{\beta\sigma_\xi^2 - \sigma_\varepsilon^2}{\sigma_\xi^2 + \sigma_\varepsilon^2}. \tag{E.5}$$

Thus the suggested regression analysis of Z_i on X_i needs modification unless $\sigma_\varepsilon^2 \ll \sigma_\xi^2$.

The interpretation of σ_ε^2 is best seen via the variance of $2\sigma_\varepsilon^2$ between two measurements on the same individual spaced in time as are the data values but with no intervening treatment: pure measurement error, thought to be negligible in the present instance, is only one component.

The representation outlined here has six parameters, namely μ, Δ, β, σ_ξ^2, σ_η^2, σ_ε^2, whereas the first and second moments of the joint distribution of X_i and Y_i give only five estimates. This points to the desirability in such studies of obtaining an independent estimate of σ_ε^2: in the analysis that follows we have assumed σ_ε^2 to be negligible. Of course, the representation could be made more complicated in various ways.

The final general point illustrated by the example concerns the availability of both systolic and diastolic measurements. With just two types of response, formal multivariate techniques for reducing dimensionality are uncalled for, although with many types of measurement such techniques could be considered. It is, however, sensible to look at the strength of relationships between the systolic and diastolic responses and this we do just by plotting the systolic difference versus the diastolic difference.

Table E.2. Analysis of difference in blood pressure (mm Hg) 2 hr after taking captopril minus value before

	Systolic	Diastolic
Mean	−18.93	−9.27
St. deviation	9.03	8.61
St. error of mean	2.33	2.22
Regr. of diff. on 'before'	−0.1233	−0.1196
St. error	0.117	0.226

The analysis. Table E.2 summarizes the analysis of the 15 differences 'after' minus 'before', taken separately for systolic and diastolic blood pressure. Negative values correspond to a lowering of blood pressure after taking captopril. Clearly there is overwhelming evidence that the mean blood pressures are lower after treatment: in the absence of a concurrent control group, interpretation is open to some ambiguity.

The regression coefficients of difference on initial value, or the corresponding plots, show no evidence that the apparent treatment effect varies with initial value, assuming that the random component of the measurements is negligibly small. It is interesting, however, that both estimated regression coefficients are negative, in line with Equation (E.5) with $\beta = 0$.

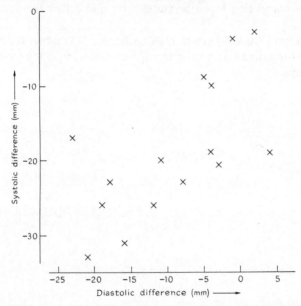

Fig. E.1. Systolic difference ('after' minus 'before') versus diastolic difference.

Figure E.1 shows the systolic difference plotted against the diastolic difference, therefore also showing implicitly the marginal distributions of the two separate differences. Two patients, namely 7 and 13, have average systolic differences but respectively large and small diastolic differences; otherwise there is quite a strong relationship. It is not known whether there is anything anomalous about the two patients or, for example, whether one of the component measurements has a gross error.

Further points and exercises.

(i) It is suggested that instead of analysing differences, 'after' minus 'before', ratios 'after' divided by 'before' should be calculated and in particular the null hypothesis that the mean ratio is 1 should be tested. Criticize the proposal and suggest an alternative method of analysis for use when ratios are judged relevant.

(ii) Plot the 'after' values against the 'before' values with systolic and diastolic measurements on the same graph. What can be concluded? What further plots might elucidate connections between the treatment effects for the two types of blood pressure?

(iii) Estimate the parameters in the model (E.1)–(E.4) assuming known values, e.g. 0, $\frac{1}{2}$, 1, 2 for the ratio $\sigma_\epsilon/\sigma_\xi$. Estimate them also assuming that $\beta = 0$.

(iv) What considerations enter in choosing between differences and other measures for comparing blood pressures after and before treatment?

Related reference. Anderson *et al.* (1980, Chapter 12) discuss the analysis of pre- and post-treatment data in comparative studies involving treatment and control groups.

Example F   Comparison of industrial processes in the presence of trend

Description of data. * In a plant-scale experiment on the production of a certain chemical, a batch of intermediate product was divided into six equal portions which were then processed on successive days by two different methods, P_1 and P_2. The order of treatment and the yields are given in Table F.1. It was expected that superposed on any process effect there would be a smooth, roughly parabolic trend. Experience of similar experiments showed that the standard deviation of a single observation was about 0.1.

Table F.1. Treatment and yields in plant-scale experiment

Day	1	2	3	4	5	6
Process	P_1	P_2	P_2	P_1	P_1	P_2
Yield	5.84	5.73	7.30	10.46	9.71	5.91

General considerations. This example illustrates in rather extreme form the fitting of a small number of observations by a model containing nearly as many parameters as there are observations. Normally this is to be avoided, exceptions being when the data are of high quality and the model fitted has fairly firm justification in previous experience or theory.

In the present instance with clear curvature and process difference to be represented, at least four parameters are inevitable. Two degrees of freedom remain. An estimate of error based on two degrees of freedom is for several reasons virtually useless on its own: the mean square associated with these two degrees of freedom is, however, important in providing a general check on the adequacy of the model, by comparison with the measure of error externally available. More broadly, in many investigations there is at least some external knowledge of the variability to be expected. Even if the investigation has a viable estimate of error on its own, as is certainly desirable, at least informal comparison with the external estimate of error is a good idea, as a check on technique.

A final general point concerns parameterization. If the calculations are done on a pocket calculator, or if some theoretical study of the analysis is

---

* Fictitious data based on a real investigation.

undertaken, parameterization to achieve near orthogonality is desirable and this is illustrated below.

The analysis. The data are plotted in Fig. F.1. Interpreted in conjunction with the external value of standard deviation the data show a clear difference between processes and, for each process, a curved trend with time. It would be possible to fit separate parabolas for each process. this involving six parameters for six observations.

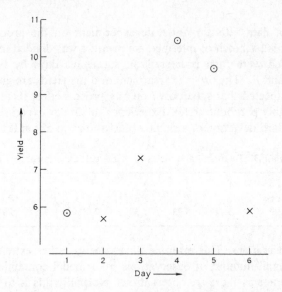

Fig. F.1. Yield in plant-scale experiment. Standard deviation for an observation equals 0.1.

⊙ Process P_1
✕ Process P_2

It seems, however, preferable to assume a common trend, i.e. parallel curves for the two processes. Equivalently, if Y_t repeats the yield at time t ($t = 1, \ldots, 6$), we are led to the model

$$E(Y_t) = \text{mean} + \text{process effect} + \text{trend},$$

where the trend is quadratic. This can be parameterized in various ways, in the end equivalent. Discussion is much simplified, however, by achieving near orthogonality and there are two aspects to this:

  (a) the process means (ignoring trend) should be written as, say, $\mu + \tau$ and $\mu - \tau$, rather than as, say, ν and $\nu + \Delta$;
  (b) orthogonal polynomials should be used for the trend.

Thus we take the model in the form

$$E(Y_t) = \begin{cases} \mu+\tau+\beta_1\xi_{1t}+\beta_2\xi_{2t} & \text{for Process 1,} \\ \mu-\tau+\beta_1\xi_{1t}+\beta_2\xi_{2t} & \text{for Process 2,} \end{cases} \tag{F.1}$$

where ξ_{1t} and ξ_{2t} represent linear and quadratic orthogonal polynomials for which the values at $t = 1, \ldots, 6$ are $\xi_{1t} = -5, -3, -1, 1, 3, 5$ and $\xi_{2t} = 5, -1, -4, -4, -1, 5$ (Pearson and Hartley, 1966, Table 47).

The least-squares equations are

$$\begin{pmatrix} 6 & 0 & 0 & 0 \\ 0 & 6 & -2 & 0 \\ 0 & -2 & 70 & 0 \\ 0 & 0 & 0 & 84 \end{pmatrix} \begin{pmatrix} \hat{\mu} \\ \hat{\tau} \\ \hat{\beta}_1 \\ \hat{\beta}_2 \end{pmatrix} = \begin{pmatrix} 44.95 \\ 7.07 \\ 15.45 \\ -27.73 \end{pmatrix} \tag{F.2}$$

with solution

$$\hat{\mu} = 44.95/6 = 7.4917,$$

$$\hat{\beta}_2 = -27.73/84 = -0.3301 \tag{F.3}$$

and

$$\begin{pmatrix} \hat{\tau} \\ \hat{\beta}_1 \end{pmatrix} = \begin{pmatrix} 6 & -2 \\ -2 & 70 \end{pmatrix}^{-1} \begin{pmatrix} 7.07 \\ 15.45 \end{pmatrix} = \begin{pmatrix} 1.2639 \\ 0.2568 \end{pmatrix}.$$

The residual sum of squares is 0.0436. Note that if this is found by the usual method of subtracting the sum of squares due to regression from the sum of squares of the original observations, care is needed to avoid serious rounding errors. Thus the residual mean square is 0.0218, which, with 2 degrees of freedom, is in entirely satisfactory agreement with the external variance of 0.01; a formal test based on $\chi_2^2 = 0.0436/0.01$ is unnecessary in this case. Thus there is no indication that a more complex model is called for.

Residuals and fitted values could be calculated and the fitted parabolas drawn on Fig. F.1. This is largely superfluous, however. Because the sum of squares of 6 residuals is 0.0436, no single residual can much exceed 0.2; also because four parameters have been fitted to six observations the residuals are highly dependent.

Variances of the fitted parameters are obtained from the inverse of the matrix in Equation (F.2), i.e.

$$\text{var}(\hat{\mu}) = \sigma^2/6, \qquad \text{var}(\hat{\tau}) = 70\sigma^2/416,$$
$$\text{var}(\hat{\beta}_1) = 6\sigma^2/416, \qquad \text{var}(\hat{\beta}_2) = \sigma^2/84,$$

where $\sigma = 0.1$.

The estimated process effect is $2\hat{\tau} = 2.528$ with standard error 0.082.

To determine the position of maximum yield we need the explicit forms of the linear and quadratic trends in Equation (F.1). These can be obtained either from first principles or from Pearson and Hartley (1966, Table 47). Thus, the fitted model is

$$\hat{Y}_t = \hat{\mu} \pm \hat{\tau} + \hat{\beta}_1(2t-7) + \hat{\beta}_2(28-21t+3t^2)/2$$

and the estimated time delay $\hat{t}$ corresponding to maximum yield is given by $\partial \hat{Y}_t/\partial t = 0$, i.e.

$$\hat{t} = 3.5 - (2\hat{\beta}_1)/(3\hat{\beta}_2)$$
$$= 4.02 \text{ days}.$$

Confidence limits for $\hat{t}$ can be calculated from Fieller's theorem, or approximately as follows:

$$\text{var}(\hat{t}) = \frac{4}{9}\,\text{var}\!\left(\frac{\hat{\beta}_1}{\hat{\beta}_2}\right)$$

$$\simeq \frac{4}{9}\{\beta_2^2\,\text{var}(\hat{\beta}_1) + \beta_1^2\,\text{var}(\hat{\beta}_2)\}/\beta_2^4. \tag{F.4}$$

Substitution of the estimated values $\hat{\beta}_1$ and $\hat{\beta}_2$ into Equation (F.4) gives the approximate standard error of $\hat{t}$ equal to 0.030, and approximate 95 per cent confidence limits for the true time delay of 3.96 to 4.08 days.

Further points and exercises

(i) What would it have been reasonable to conclude from the data had the external value of standard deviation not been available?

(ii) Discuss the partition of the two degrees of freedom for residual into parts, in particular to assess any difference in curvature between processes, i.e. process × curvature interaction.

(iii) Comment on the particular arrangement of treatments used. What general principle is involved in the choice of this arrangement?

(iv) Verify that if the residual sum of squares is computed in the usual way but using least-squares estimates correct to only three decimal places, there results a rounding error of approximately a factor of two. Discuss the general implications.

Example G

Cost of construction of nuclear power plants

Description of data. Table G.1 gives data, reproduced by permission of the Rand Corporation, from a report (Mooz, 1978) on 32 light water reactor (LWR) power plants constructed in the USA. It is required to predict the capital cost involved in the construction of further LWR power plants. The notation used in Table G.1 is explained in Table G.2. The final six lines of data in Table G.1 relate to power plants for which there were partial turnkey guarantees and for which it is possible that some manufacturers' subsidies might be hidden in the quoted capital costs.

General considerations. One of the most common problems in advanced applied statistics is the study of the relation between a single continuous response variable and a number of explanatory variables. When the expected response can be represented as a linear combination of unknown parameters, with co-efficients determined by the explanatory variables, and when the error structure is suitably simple, the techniques of multiple regression based on the method of least squares are applicable. The formal theory of multiple regression, and the associated significance tests and confidence regions, have been extensively developed and are described in numerous textbooks; see, for example, Draper and Smith (1981) and Seber (1977). Further, computer programs for implementing the methods are widely available.

Nevertheless, there can be difficulties, partly of technique but more importantly of interpretation, in applying the methods, especially to observational data with fairly large numbers of explanatory variables. We now mention briefly some commonly occurring points. Of course, in any particular application many of the potential difficulties may be absent and indeed the present example seems relatively well behaved.

Some issues that arise fairly commonly are the following:

(i) What is the right general form of model to fit?
(ii) Are there aspects of error structure that seriously affect the analysis?
(iii) Are there outliers or anomalous observations that need to be isolated?
(iv) What can be done if a subset of observations is isolated, possibly not following the same model as the main body of data?

81

Table G.1. Data on thirty-two LWR power plants in the USA

C	D	T_1	T_2	S	PR	NE	CT	BW	N	PT
460.05	68.58	14	46	687	0	1	0	0	14	0
452.99	67.33	10	73	1065	0	0	1	0	1	0
443.22	67.33	10	85	1065	1	0	1	0	1	0
652.32	68.00	11	67	1065	0	1	1	0	12	0
642.23	68.00	11	78	1065	1	1	1	0	12	0
345.39	67.92	13	51	514	0	1	1	0	3	0
272.37	68.17	12	50	822	0	0	0	0	5	0
317.21	68.42	14	59	457	0	0	0	0	1	0
457.12	68.42	15	55	822	1	0	0	0	5	0
690.19	68.33	12	71	792	0	1	1	1	2	0
350.63	68.58	12	64	560	0	0	0	0	3	0
402.59	68.75	13	47	790	0	1	0	0	6	0
412.18	68.42	15	62	530	0	0	1	0	2	0
495.58	68.92	17	52	1050	0	0	0	0	7	0
394.36	68.92	13	65	850	0	0	0	1	16	0
423.32	68.42	11	67	778	0	0	0	0	3	0
712.27	69.50	18	60	845	0	1	0	0	17	0
289.66	68.42	15	76	530	1	0	1	0	2	0
881.24	69.17	15	67	1090	0	0	0	0	1	0
490.88	68.92	16	59	1050	1	0	0	0	8	0
567.79	68.75	11	70	913	0	0	1	1	15	0
665.99	70.92	22	57	828	1	1	0	0	20	0
621.45	69.67	16	59	786	0	0	1	0	18	0
608.80	70.08	19	58	821	1	0	0	0	3	0
473.64	70.42	19	44	538	0	0	1	0	19	0
697.14	71.08	20	57	1130	0	0	1	0	21	0
207.51	67.25	13	63	745	0	0	0	0	8	1
288.48	67.17	9	48	821	0	0	1	0	7	1
284.88	67.83	12	63	886	0	0	0	1	11	1
280.36	67.83	12	71	886	1	0	0	1	11	1
217.38	67.25	13	72	745	1	0	0	0	8	1
270.71	67.83	7	80	886	1	0	0	1	11	1

Table G.2. Notation for data of Table G.1

C	Cost in dollars $\times 10^{-6}$, adjusted to 1976 base
D	Date construction permit issued
T_1	Time between application for and issue of permit
T_2	Time between issue of operating license and construction permit
S	Power plant net capacity (MWe)
PR	Prior existence of an LWR on same site ($= 1$)
NE	Plant constructed in north-east region of USA ($= 1$)
CT	Use of cooling tower ($= 1$)
BW	Nuclear steam supply system manufactured by Babcock–Wilcox ($= 1$)
N	Cumulative number of power plants constructed by each architect–engineer
PT	Partial turnkey plant ($= 1$)

(v) Is it feasible to simplify the model, normally by reducing the number of explanatory variables?

(vi) What are the limitations on the interpretation and application of the final relation achieved?

All these points, except the key issue (vi), can to some extent be dealt with formally, for instance by comparing the fits of numerous competing models. Often, though, this would be a ponderous way to proceed.

Consideration of point (i), choice of form of relation, involves a possible transformation of response variable, in the present instance cost and log cost being two natural variables for analysis, and a choice of the nature and form of the explanatory variables. For instance, should the explanatory variables, where quantitative, be transformed? Should derived explanatory variables be formed to investigate interactions? Frequently in practice, any transformations are settled on the basis of general experience: the need for interaction terms may be examined graphically or, especially with large numbers of explanatory variables, may be checked by looking only for interactions between variables having large 'main effects'. In the present example, log cost has been taken as response variable and the explanatory variables S, T_1, T_2 and N have also been taken in log form, partly to lead to unit-free parameters whose values can be interpreted in terms of power-law relations between the original variables. It is plausible that random variations in cost should increase with the value of cost and this is another reason for the log transformation.

Complexities of error structure, point (ii), can arise via systematic changes in variance, via notable non-normality of distribution and, particularly importantly, via correlation in errors for different individuals. All these effects may be of intrinsic interest, but more commonly have to be considered either because a modification of the method of least squares is called for or because, while the least-squares estimates and fit may be satisfactory, the precision of the least-squares estimates may be different from that indicated under standard assumptions. In particular, substantial presence of positive correlations can mean that the least-squares estimates are much less precise than standard formulae suggest. A special form of correlated error structure is that of clustering of individuals into groups, the regression relations between and within groups being quite different. There is no sign that any of these complications are important in the present instance.

Somewhat related is point (iii), occurrence of outliers. Where interest is focused on particular regression coefficients, the most satisfactory approach is to examine informally or formally whether there is any single observation or small set of observations whose omission would greatly change the estimate in question; see also point (iv).

In the present example there is a group of 6 observations distinct from the main body of 26 and there is some doubt whether the 6 should be included.

This is quite a common situation; the possibly anomalous group may, for example, have extreme values of certain explanatory variables. The most systematic approach is to fit additional linear models to test consistency. Thus one extra parameter can be fitted to allow for a constant displacement of the anomalous group and the significance of the resulting estimate tested. A more searching analysis is provided by allowing the regression coefficients also to be different in the anomalous group; in the present instance this has been done one variable at a time, because with 10 explanatory variables and only 6 observations in the separate group there are insufficient observations to examine for anomalous slopes simultaneously.

Point (v), the simplification of the fitted model, is particularly important when the number of explanatory variables is large, and even more particularly when there is an *a priori* suspicion that many of the explanatory variables are measuring essentially equivalent things. The need for such simplification arises particularly, although by no means exclusively, in observational studies. More explicitly, the reasons for seeking simplification are that:

(a) estimation of parameters of clear interest can be degraded by including unnecessary terms in the model:

(b) prediction of response of new individuals is less precise if unnecessary terms are included in the predictor;

(c) it is often reasonable to expect that when many explanatory variables are available only a few will have a major effect on response and it may be of primary interest to isolate these important variables and to interpret their effects;

(d) it may be desirable to simplify future work by recording only a smaller number of explanatory variables.

Techniques for the retention of variables are, as explained in Section 3.4 of Part I, forward, backward or some mixture. Where some of the parameters represent effects of direct interest they should be included regardless of the operation of a selection procedure. It is entirely possible that forward selection leads to a different equation from backward selection, although this has not happened in the present example. It is therefore important, especially where interpretation of the particular form of equation is central to the analysis, that if there are several simple equations that fit almost equally well, all should be isolated for consideration and not one chosen somewhat arbitrarily.

Suppose now that a representation, hopefully quite a simple one, has been obtained for expected response as a function of certain explanatory variables. What are the principal aspects in using and interpreting such an equation? This is point (vi) of the list above. There are at least five rather different possibilities.

Firstly, an equation such as that summarized in Table G.4, including the residual standard deviation, provides a concise description of the data, as regards the dependence of cost on the other variables. Such a description can be useful in thinking about the data qualitatively and in comparing different, somewhat related, sets of data.

A second descriptive use is in the study of the individual cases. The residual from the fitted model is an index for each power station assessing its cost relative to what might have been anticipated given the explanatory variables.

Thirdly, the equation can be used for prediction. A new individual has given (or sometimes predicted) values of the relevant explanatory variables and the equation, and associated measures of variability, are used to forecast cost, preferably with confidence limits. In such prediction the main assumption, in addition to the technical adequacy of the model in the region of explanatory variables required for prediction, is that any unmeasured variable affecting response keeps the same statistical relationship with the measured explanatory variables as obtains in the data. Thus, in particular, if the new individual to be predicted differs in some way from the reference data, other than is directly or indirectly accounted for in the explanatory variables, a modification of the regression predictor is worth consideration. For example, a major technological innovation between the data analysed and the individual to be predicted would call for such modification of the predictor.

Fourthly, the equation may be used to predict for a new individual, or sometimes for one of the original individuals, the consequences of changes in one or more of the explanatory variables. For example, one might wish to predict not so much the cost for a new individual as the change in cost for that individual as size changes. The relevant regression coefficient predicts that change, provided that the other explanatory variables are held fixed and that any important unobserved explanatory variables change appropriately with the change in size. The prediction of changes in uncontrolled observational systems, e.g. in the social sciences, needs particularly careful specification of the changes in explanatory variables envisaged.

Finally, and in some ways most importantly, one may hope to gain insight into the system under study by careful inspection of which explanatory variables contribute appreciably to the response and of the signs and magnitudes of the associated regression coefficients. Thus in the present example, why do certain variables appear not to contribute appreciably, why is the regression coefficient on log size appreciably less than one, the value for proportionality, and so on? As indicated in the previous paragraph, the regression coefficients estimate changes in response under perturbations of the system whose precise specification needs care.

The last two applications of regressions need considerable thought, especially if there is any possibility that an important explanatory variable has been overlooked.

The analysis. As explained in the preceding section, we take $\log C$, $\log S$, $\log N$, $\log T_1$ and $\log T_2$; throughout natural logs are used.

A regression of log C on all ten explanatory variables gives a residual mean square of $0.5680/21 = 0.0270$ with 21 degrees of freedom. Elimination of non-significant variables successively one at a time removes BW, log T_1, log T_2 and PR (Table G.3), leaving six variables and a residual mean square of $0.6337/25 = 0.0253$ with 25 degrees of freedom; the residual standard deviation is 0.159.

Table G.3. Elimination of variables

No. variables included	Variables eliminated	Residual	
		s.s.	d.f.
10	—	0.5680	21
9	BW	0.5709	22
8	log T_1	0.5730	23
7	log T_2	0.6165	24
6	PR	0.6337	25

None of the eliminated variables is significant if re-introduced. The estimated coefficients and standard errors for the six-variable regression are given in Table G.4. The variable PT, denoting partial turnkey guarantee, has a co-efficient of -0.2261, with a standard error of 0.1135 (25 d.f.), suggesting that cost tends to be reduced on average by about 20 per cent for these six plants.

Table G.4. Multiple regression: full and reduced models

Variable	Regression coefficient				
	Reduced model			Full model	
	Estimate	Standard error		Estimate	Standard error
Constant	-13.26	3.140		-14.24	4.229
PT	-0.2261	0.1135		-0.2243	0.1225
CT	0.1404	0.0604		0.1204	0.0663
log N	-0.0876	0.0415		-0.0802	0.0460
log S	0.7234	0.1188		0.6937	0.1361
D	0.2124	0.0433		0.2092	0.0653
NE	0.2490	0.0741		0.2581	0.0769
log T_1	—	—		0.0919	0.2440
log T_2	—	—		0.2855	0.2729
PR	—	—		-0.0924	0.0773
BW	—	—		0.0330	0.1011
Residual st. dev.	0.159 (25 d.f.)			0.164 (21 d.f.)	

To check whether these six plants and the twenty-six others can be fitted by a model with common coefficients for each of the variables CT, log N, log S and log D, we include in turn in the regression the interaction of each variable with PT. This cannot be done for the variable NE since all six PT plants were constructed in the same region. Table G.5 summarizes the results. None of the interaction coefficients is significant. We note that the coefficients of the six

Table G.5. Regressions including interactions with PT

Variable	$Z = CT$ Estimate	s.e.	$Z = \log N$ Estimate	s.e.	$Z = \log S$ Estimate	s.e.	$Z = D$ Estimate	s.e.
Constant	−13.23	3.193	−13.26	3.225	−13.09	3.239	−13.22	3.231
PT	−0.2429	0.1221	−0.2293	0.8265	−2.188	5.854	−1.529	15.17
CT	0.1312	0.0652	0.1404	0.0629	0.1400	0.0615	0.1412	0.0624
log N	−0.0868	0.0422	−0.0876	0.0423	−0.0868	0.0423	−0.0875	0.0423
log S	0.7229	0.1208	0.7234	0.1219	0.7176	0.1222	0.7222	0.1221
D	0.2121	0.0440	0.2124	0.0444	0.2104	0.0444	0.2120	0.0444
NE	0.2490	0.0754	0.2490	0.0757	0.2484	0.0755	0.2489	0.0757
PT $\times Z$	0.0798	0.1887	0.0014	0.3683	0.2916	0.8700	0.0193	0.2246

common variables in Table G.5 remain fairly stable, except for PT which, in two cases, is estimated very imprecisely. A model with common coefficients as given in Table G.4 seems reasonable. With this model the predicted cost increases with size, although less rapidly than proportionally to size, is further increased if a cooling tower is used or if constructed in the NE region, but decreases with experience of architect–engineer.

Table G.6. Comparison of observed and fitted values based on six-variable regression of Table G.4 fitted to log C

Observed	Fitted	Residual	Observed	Fitted	Residual
6.131	6.051	0.080	6.568	6.379	0.189
6.116	6.225	−0.109	5.669	5.891	−0.222
6.094	6.225	−0.131	6.781	6.492	0.289
6.481	6.398	0.083	6.196	6.230	−0.034
6.465	6.398	0.067	6.342	6.178	0.164
5.845	5.976	−0.131	6.501	6.651	−0.150
5.607	5.934	−0.327	6.432	6.249	0.183
5.760	5.704	0.056	6.411	6.384	0.027
6.125	5.987	0.138	6.160	6.129	0.031
6.537	6.411	0.126	6.547	6.797	−0.250
5.860	5.789	0.071	5.335	5.401	−0.066
5.998	6.262	−0.264	5.665	5.606	0.059
6.021	5.891	0.130	5.652	5.621	0.031
6.206	6.241	−0.035	5.636	5.621	0.015
5.977	6.016	−0.039	5.382	5.401	−0.019
6.048	5.992	0.056	5.601	5.621	−0.020

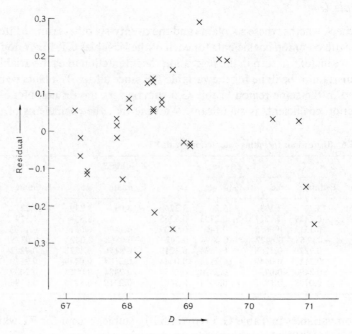

Fig. G.1. Residuals of log *C* from six-variable model versus *D*, date construction permit issued.

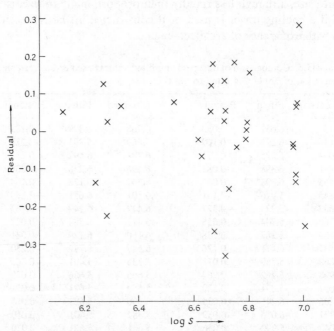

Fig. G.2. Residuals of log *C* from six-variable model versus log *S*, power plant net capacity.

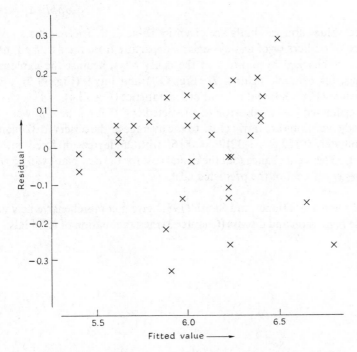

Fig. G.3. Residuals of log C from six-variable model versus fitted values.

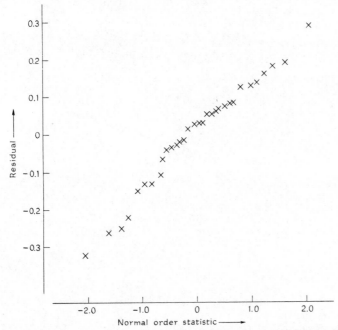

Fig. G.4. Residuals of log C from six-variable model versus normal order statistics.

Fitted values and residuals are given in Table G.6. The residuals give no evidence of outliers or of any systematic departure from the assumed model; this can be checked by plotting in the usual ways, against the explanatory variables, for example against D (Fig. G.1) and log S (Fig. G.2), against fitted values (Fig. G.3) and normal order statistics (Fig. G.4).

The estimated standard error of predicted log C for a new power plant, provided conditions are fairly close to the average of the observed 32 plants, is approximately $0.159 (1 + 1/32)^{1/2} = 0.161$ with 25 degrees of freedom. Thus there is a 95 per cent chance that the actual cost for the new plant will lie within about ± 39 per cent of the predicted cost.

Related reference. Draper and Smith (1981) give a comprehensive account of multiple regression and discuss (Chapter 3) the examination of residuals.

Example H

Effect of process and purity index on fault occurrence

Description of data.* Minor faults occur irregularly in an industrial process and, as an aid to their diagnosis, the following experiment was done. Batches of raw material were selected and each batch was divided into two equal sections: for each batch, one of the sections was processed by the standard method and the other by a slightly modified process, in which the temperature at one stage is reduced. Before processing, a purity index was measured for the whole batch of material. For the product from each section of material it was recorded whether the minor faults did or did not occur. Results for 22 batches are given in Table H.1.

Table H.1. Occurrence of faults in 22 batches

Purity index	Standard process	Modified process	Purity index	Standard process	Modified process
7.2	NF	NF	6.5	NF	F
6.3	F	NF	4.9	F	F
8.5	F	NF	5.3	F	NF
7.1	NF	F	7.1	NF	F
8.2	F	NF	8.4	F	NF
4.6	F	NF	8.5	NF	F
8.5	NF	NF	6.6	F	NF
6.9	F	F	9.1	NF	NF
8.0	NF	NF	7.1	F	NF
8.0	F	NF	7.5	NF	F
9.1	NF	NF	8.3	NF	NF

F, Faults occur.   NF, No faults occur.

General considerations. The data here are so limited that in practice very detailed analysis would hardly be justified. The unusual features of the data are the pairing, combined with the availability of a quantitative explanatory variable; the response variable is binary.

Rather than plunge straight into the fitting of relatively complex models, it is wise to start by simple analysis, first ignoring the explanatory variable and then

* Fictitious data based on a real investigation.

Applied statistics

examining the effect of that variable by simple graphs or tables. Maximum-likelihood fitting of various models can then follow, with a simple basis having been laid for understanding the answers.

The analysis. If we ignore purity index, a standard technique for assessing matched-pair data with binary responses involves the 14 pairs with mixed response, these being split between 5 'NF, F' and 9 'F, NF'. This suggests a higher chance of fault in the standard process. The null hypothesis of absence of process difference is tested via the binomial distribution with 14 trials and probability $\frac{1}{2}$; the two-sided level obtained via a normal approximation with continuity correction is

$$2\Phi\left(-\frac{|5-7|-\frac{1}{2}}{\sqrt{(14\times\frac{1}{2}\times\frac{1}{2})}}\right) = 2\Phi(-0.802) = 0.423, \tag{H.1}$$

so that the apparent process effect is entirely consistent with chance fluctuations.

To examine the effect of purity index on its own, a graph, Fig. H.1, of grouped proportion of faults versus purity index shows that most of the faults occur on batches of low purity index.

To investigate both effects, we fit a linear logistic model by maximum likelihood. To fit from first principles, rather than by a package such as GLIM, the model is taken in the approximately orthogonal form that for the ith batch with purity index x_i,

$$\text{pr(fault|standard process)} = \frac{\exp[\alpha+\Delta+\beta(x_i-\bar{x})]}{1+\exp[\alpha+\Delta+\beta(x_i-\bar{x})]}$$

$$\tag{H.2}$$

$$\text{pr(fault|modified process)} = \frac{\exp[\alpha-\Delta+\beta(x_i-\bar{x})]}{1+\exp[\alpha-\Delta+\beta(x_i-\bar{x})]},$$

where $\bar{x}$ is the mean of the x_i. It is provisionally assumed that responses are independent when purity index is in the model.

Table H.2 compares the results of fitting the model (H.2) and of reduced models with $\Delta = 0$, with $\beta = 0$ and with $\beta = \Delta = 0$. The last two models correspond to two and one simple binomial distributions. The results confirm the simpler analyses. The data are not consistent with constant probability of fault; χ^2 with 2 degrees of freedom is

$$2(-26.406+29.767) = 6.72$$

and is just significant at 5 per cent. The estimate of the logistic process difference 2Δ is 0.864 with a standard error of 0.672. Thus a wide range of differences, including $\Delta = 0$, is consistent with the data. Positive means that the standard

Table H.2. Fitting of linear logistic models

Model	No. of parameters	Max. log lik.	Estimates and standard errors	
Mean	3	−26.406	$\hat{\alpha}$	−0.412±0.333
Process difference			$\hat{\Delta}$	0.432±0.336
Purity index			$\hat{\beta}$	−0.604±0.284
Mean	2	−29.010	$\hat{\alpha}$	−0.381±0.313
Process difference			$\hat{\Delta}$	0.381±0.313
Mean	2	−27.261	$\hat{\alpha}$	−0.395±0.461
Purity index			$\hat{\beta}$	−0.579±0.391
Mean	1	−29.767	$\hat{\alpha}$	−0.368±0.307

process has the higher probability of fault. A trend with purity index is moderately well established; $\hat{\beta} = -0.604$, with a standard error of 0.284, the two-sided P value is less than 5 per cent and the trend is in the direction expected on general grounds.

To interpret the parameters and to check the adequacy of the model, Fig. H.1, shows the fitted models, i.e. the curves (H.2) with α, β, Δ replaced by $\hat{\alpha}$, $\hat{\beta}$, $\hat{\Delta}$, The figure also shows the observed proportions of faults based on a grouping into five sets with roughly constant purity index in each set. The plot exposes the paucity of the data, revealed of course also by very wide confidence limits for β and Δ; if desired these limits, too, could be illustrated graphically.

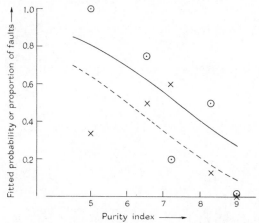

Fig. H.1. Fitted logistic models ———— Standard process
— — — — — Modified process
Observed proportion of faults ⊙ Standard process
✕ Modified process

Further points and exercises.

(i) Find the exact binomial probability corresponding to Equation (H.1).

(ii) Compare tests of $\beta = 0$ and of $\Delta = 0$ from maximum-likelihood estimates and their standard error, with those based on maximized log likelihood.

(iii) The model (H.2) assumes independence of the two responses in a pair. How can this be tested and, if necessary, dependence allowed for?

(iv) An alternative to the linear logistic model is to record 1 for fault, 0 for no fault, to fit a linear model for expected response, i.e. for the probability of a fault. Compare this with the results of Table H.2; under what circumstances would this approach be expected to give appreciably different answers from the linear logistic model?

Related references. The comparison of binary data in matched pairs is described by Armitage (1971, §16.2) and Wetherill (1967, p. 193). Linear logistic models are discussed by Armitage (1971, §12.5), Cox (1970), and Snedecor and Cochran (1967, §§16.8—16.12). Computer programs for their fitting by maximum likelihood are fairly widely available.

Example I

Growth of bones from chick embryos

Description of data. Table I.1 gives data on the growth of bones from seven-day-old chick embryos after cultivation over a nutrient chemical medium (Biggers and Heyner, 1961). The observations are of log dry weight (μg). Two bones were available from each embryo and the experiment was therefore set out in a (balanced) incomplete block design with two units per block. C denotes the complete medium with about 30 ingredients in carefully controlled quantities. The five other media were obtained by omitting a single amino acid, e.g., His$^-$ is a medium without L-histidine, etc. The treatment pairs were randomized, but the following results are given in systematic order. Interest lies in comparing the effects of omitting the various amino acids.

Table I.1. Log$_{10}$(dry weight) of tibiotarsi from seven-day-old chick embryos

Embryo	1	C	2.51: His$^-$	2.15		9	His$^-$	2.32: Lys$^-$	2.53
	2	C	2.49: Arg$^-$	2.23		10	Arg$^-$	2.15: Thr$^-$	2.23
	3	C	2.54: Thr$^-$	2.26		11	Arg$^-$	2.34: Val$^-$	2.15
	4	C	2.58: Val$^-$	2.15		12	Arg$^-$	2.30: Lys$^-$	2.49
	5	C	2.65: Lys$^-$	2.41		13	Thr$^-$	2.20: Val$^-$	2.18
	6	His$^-$	2.11: Arg$^-$	1.90		14	Thr$^-$	2.26: Lys$^-$	2.43
	7	His$^-$	2.28: Thr$^-$	2.11		15	Val$^-$	2.28: Lys$^-$	2.56
	8	His$^-$	2.15: Val$^-$	1.70					

General considerations. This experiment forms a balanced incomplete block design. It would be possible to analyse it via the special formulae for such designs (Cochran and Cox, 1957, §11.5) but, partly because of the special simplicity resulting from the use of only two treatments per 'block', it is more instructive to proceed from first principles.

The simplest analysis of such designs, the so-called within-block analysis, regards variation between pairs as arbitrary, each pair thus in effect having an associated parameter unique to that pair. The only way that conclusions about the treatments can be drawn free of these pair parameters is in effect by forming a single derived response for each pair, eliminating pair effects. This is most simply done by taking the difference between the two values for a pair as a derived response having expectation the difference between the corresponding

95

true effects for the two treatments concerned. This implicitly makes the plausible assumption that the effect of omitting a single amino acid is to sub- tract from log weight a characteristic (unknown) amount, i.e. that weight itself is reduced by some unknown fraction. This leads to a linear model for the derived responses and hence to the application of the method of least squares.

For incomplete block designs like this, a second independent analysis can be made in which the pair parameters are regarded as random and the pair sums or means regarded as a second derived response variable. This leads to a new set of least-squares estimates, independent of the first set, and the final estimates are appropriately weighted means of the within-pair and between-pair esti- mates. It is quite common practice to omit the second analysis and this is sensible wherever there is substantial variation between blocks, although when the block size is small, and especially as here where the block size is two, up to half the information may be in the between-block analysis.

The analysis. If Y_{ij} denotes the jth observation ($j = 1, 2$) on the ith block ($i = 1, \ldots, 15$), we base the analysis on the linear model

$$Y_{ij} = \mu + \beta_i + \tau_k + \varepsilon_{ij}, \qquad (I.1)$$

where μ represents an overall constant, β_i a block parameter, τ_k the effect due to treatment k and ε_{ij} is random error.

For the within-block analysis we consider the differences within blocks, $W_i = Y_{i_1} - Y_{i_2}$, with expected value

$$E(W_i) = \tau_{k_{1i}} - \tau_{k_{2i}}, \qquad (I.2)$$

k_{1i} and k_{2i} being the two treatments applied to block i. More conveniently, we can write

$$\theta_k = \tau_C - \tau_k, \qquad (I.3)$$

where τ_C denotes the effect due to the complete medium, i.e. θ_k ($k = 1, \ldots, 5$) represents the effect of omitting the kth amino acid; the θ_k are the quantities directly of interest. The model is thus

$$E(W) = X\theta \qquad (I.4)$$

with $W^T = (W_1, \ldots, W_{15})$ and $\theta^T = (\theta_1, \ldots, \theta_5)$ and X is a simple 15×5 matrix with elements $1, 0$ and -1. The least-squares solution is straightforward and the estimates $\hat{\theta}_k$ are given in Table I.2. The residual mean square is 0.0132 with 10 degrees of freedom and the estimated standard error of $\hat{\theta}_k$ is 0.066.

For the between-block analysis we assume β_i in Equation (I.1) to be a random variable with $E(\beta_i) = 0$ and analyse the block totals $Z_i = Y_{i1} + Y_{i2}$. From Equation (I.1),

$$E(Z_i) = 2\mu + \tau_{k_{1i}} + \tau_{k_{2i}}, \qquad (I.5)$$

Example I 97

Table I.2. Estimates of parameters θ_k

	Amino acid				
	His $k = 1$	Arg $k = 2$	Thr $k = 3$	Val $k = 4$	Lys $k = 5$
Within-block estimate, $\hat{\theta}_k$	0.218	0.353	0.348	0.490	0.160
Between-block estimate, $\tilde{\theta}_k$	0.553	0.395	0.333	0.420	−0.065
Pooled estimate, θ_k^*	0.286	0.361	0.345	0.476	0.115
	Est. s.e. $(\hat{\theta}_k) = 0.066$, 10 d.f.				
	Est. s.e. $(\tilde{\theta}_k) = 0.132$, 8 d.f.				
	Est. s.e. $(\theta_k^*) = 0.059$				

which may be reparameterized in terms of θ_k's, although it is simpler in this case to fit Equation (I.5) directly, assuming $\Sigma \tau_k = 0$, and to obtain estimates $\tilde{\theta}_k$, say, by differencing the least-squares estimates $\hat{\tau}_C$ and $\hat{\tau}_k$. The estimates $\tilde{\theta}_k$ are given in Table I.2. The residual mean square from Equation (I.5) is 0.0348, which, as expected, is larger than that from the within-block analysis. The estimated variance of $\tilde{\theta}_k$ is 0.0174 with 8 degrees of freedom.

We pool the within-block estimates $\hat{\theta}_k$ and the between-block estimates $\tilde{\theta}_k$, weighting each inversely proportionally to its estimated variance, to obtain pooled estimates θ_k^*, given in Table I.2. These are very similar to the estimates from the within-block analysis. Each amino acid has a significant effect upon the weight.

The data in Table I.1 are $\log_{10}$ (dry weight). Thus after taking antilogarithms of θ_k^*, the estimated percentage reductions in weight when each of the amino acids His, Arg, Thr, Val and Lys is individually omitted are 48, 56, 55, 67 and 23 per cent respectively.

Further points and exercises.

(i) Determine confidence limits for the above estimated percentage reductions in weight. This may be done using the pooled estimate of variance and 'effective' degrees of freedom in the manner of Example S.

(ii) It would be possible to do the above analysis with weight rather than log weight as the response variable, or indeed with other functions of weight. What considerations are involved in choosing between these different analyses?

Example J   Factorial experiment on cycles to failure of worsted yarn

Description of data. In an unpublished report to the Technical Committee, International Wool Textile Organization, A. Barella and A. Sust gave the data in the first four columns of Table J.1, concerning the number of cycles to failure of lengths of worsted yarn under cycles of repeated loading. The three factors which varied over levels specified in coded form in the first three columns, are

x_1, length of test specimen	(250, 300, 350 mm);
x_2, amplitude of loading cycle	(8, 9, 10 mm);
x_3, load	(40, 45, 50 g).

General considerations. There are a number of reasons why use of log cycles to failure is likely to be the most effective way of analysing these data. Firstly, relationships of the type $y \propto x_1{}^{\beta_1} x_2{}^{\beta_2} x_3{}^{\beta_3}$ are quite commonly found in the physical sciences as reasonably close approximations to empirical behaviour. Secondly, the resulting parameters β_1, β_2 and β_3 are dimensionless and thus, especially if they are close to simple integers, relatively easy to interpret. Thirdly, provided that the signs of β_1, β_2 and β_3 are appropriate, sensible limiting behaviour as the x's tend to zero and infinity is achieved. All these points concern the form of the systematic variation.

As for the random variation, again a log transformation is likely to be sensible. Cycles to failure vary over a very wide range (by a factor of 40 in fact) and the amount of random variation is likely to increase with the mean cycles to failure. More specifically, under one of the physically simplest hypotheses the effect of changing factor levels is to multiply the 'lifetime' of a particular individual by a constant. This, sometimes called the central assumption of accelerated life testing, implies that the coefficient of variation of cycles to failure, Y, is constant and thus that the standard deviation of log Y is constant.

While these two lines of argument suggest on general grounds taking log Y as response and log x_1, log x_2 and log x_3 as explanatory variables, of course an empirical test of the suitability of this analysis is still needed.

The balanced nature of the experimental design has two closely related consequences. One is that least-squares fitting of various models representing first-degree, second-degree, etc., regression of log Y on log x_1, log x_2 and log x_3

Table J.1. Cycles to failure, transformed values, fitted values and residuals

			Cycles	Log cycles		
x_1	x_2	x_3	obs	obs	fitted	resid.
−1	−1	−1	674	6.51	6.52	−0.01
−1	−1	0	370	5.91	6.11	−0.20
−1	−1	1	292	5.68	5.74	−0.06
−1	0	−1	338	5.82	5.85	−0.03
−1	0	0	266	5.58	5.44	0.14
−1	0	1	210	5.35	5.07	0.28
−1	1	−1	170	5.14	5.26	−0.12
−1	1	0	118	4.77	4.84	−0.07
−1	1	1	90	4.50	4.48	0.02
0	−1	−1	1414	7.25	7.42	−0.17
0	−1	0	1198	7.09	7.01	0.08
0	−1	1	634	6.45	6.64	−0.19
0	0	−1	1022	6.93	6.76	0.17
0	0	0	620	6.43	6.34	0.09
0	0	1	438	6.08	5.97	0.11
0	1	−1	442	6.09	6.16	−0.07
0	1	0	332	5.81	5.75	0.06
0	1	1	220	5.39	5.38	0.01
1	−1	−1	3636	8.20	8.18	0.02
1	−1	0	3184	8.07	7.77	0.30
1	−1	1	2000	7.60	7.40	0.20
1	0	−1	1568	7.36	7.52	−0.16
1	0	0	1070	6.98	7.11	−0.13
1	0	1	566	6.34	6.74	−0.40
1	1	−1	1140	7.04	6.92	0.12
1	1	0	884	6.78	6.51	0.27
1	1	1	360	5.89	6.14	−0.25

x_1, length; x_2, amplitude of loading cycle; x_3, load.

is computationally very simple. The other is that the general form of the systematic variation can be studied very easily and directly from appropriate mean values collected in two-way and one-way tables, as for other forms of balanced factorial experiment. While the final summary of conclusions is likely to be primarily in terms of a fitted regression equation, the explanatory variables being quantitative in nature, critical inspection of two-way tables is all the same an important intermediate step in the analysis.

The analysis. Table J.2 gives two-way and marginal means of log cycles to failure; throughout natural logs are used. The two-way tables show little evidence of interaction and the marginal means show that the variation with factor levels is predominantly linear; the factor levels are not quite equally spaced in terms of log x.

Table J.2. Two-way and one-way means

Load	Amplitude of loading cycle			Load	Length of test specimen			
	−	0	+		−	0	+	Mean
−	7.32	6.70	6.09	−	5.82	6.76	7.53	6.70
0	7.02	6.33	5.79	0	5.42	6.44	7.28	6.38
+	6.58	5.92	5.26	+	5.18	5.97	6.61	5.92
Mean	6.97	6.32	5.71		5.47	6.39	7.14	

Amplitude of loading cycle	Length of test specimen			
	−	0	+	Mean
−	6.03	6.93	7.16	6.97
0	5.58	6.48	6.89	6.32
+	4.80	5.76	6.57	5.71
Mean	5.47	6.39	7.14	

Extraction of linear components of main effects, equivalent to the linear model

$$\log Y = \alpha + \beta_1 \log x_1 + \beta_2 \log x_2 + \beta_3 \log x_3 + \varepsilon, \qquad (J.1)$$

is done either by direct least-squares fitting, or equivalently by extracting the linear regression component from the marginal means. There results

$$\hat{\beta}_1 = 4.957, \qquad \hat{\beta}_2 = -5.651, \qquad \hat{\beta}_3 = 3.501. \qquad (J.2)$$

As already noted, inspection of Table J.1 shows that the model in Equation (J.1) is likely to account for most of the systematic variation. To examine this in more detail, six more degrees of freedom have been isolated, i.e. six more parameters added to Equation (J.1). These are respectively linear-by-linear interactions, i.e. product terms such as $\beta_{23} \log x_2 \log x_3$ and, pure quadratic terms such as $\beta_{11} (\log x_1)^2$, taken for convenience in a form orthogonalized with respect to the parameters in Equation (J.1).

Table J.3. Analysis of variance

	d.f.	s.s.	m.s.
Length of test specimen, x_1 (linear)	1	12.5415	
Ampl. of loading cycle, x_2 (linear)	1	7.1595	
Load, x_3 (linear)	1	2.7487	
Ampl. (linear) × load (linear)	1	0.0054	
Load (linear) × length (linear)	1	0.0529	
Length (linear) × ampl. (linear)	1	0.0211	
Length (quad.)	1	0.0013	
Ampl. (quad.)	1	0.0007	
Load (quad.)	1	0.0480	
Total second-degree terms	6	0.1294	0.02157
Residual	17	0.6368	0.03746
Total	26	23.2159	

Table J.3 gives the analysis of variance. The total contribution of quadratic terms has a mean square rather less than that for residual. It is immaterial whether the error of the estimates (J.2) is obtained via the residual mean square from the linear model (J.1) or from the residual mean square of the extended model with quadratic terms. To be slightly cautious, the second and rather larger value has been taken, giving a residual standard deviation of 0.194 and estimated standard errors for (J.2) of

$$0.271, \quad 0.409, \quad 0.409$$

with 17 degrees of freedom.

The near equality of $\hat{\beta}_1$ and $-\hat{\beta}_2$ suggests that the dependence on x_1 and x_2 can be expressed in terms of x_2/x_1, and this is particularly appealing on dimensional grounds because both x_2 and x_1 are lengths. The composite variable x_2/x_1 is the fractional extension of the loading cycle. It would, however, not be correct to argue that by dimensional analysis any dependence can only be on the dimensionless variable x_2/x_1, because there are other lengths implied in the problem, notably the mean fibre length.

The data are in fact quite closely fitted by the simple relationship

$$y \propto \left(\frac{x_2}{x_1}\right)^{-5} x_3^{-7/2}.$$

It would be instructive to compare the residual standard deviation of 0.154, corresponding to a coefficient of variation of about 20 per cent, with any value that might be available for repeat tests under the same conditions. A graph of the residuals versus fitted values gives no evidence that the error of log cycles to failure varies with the mean response: thus the data seem reasonably consistent with the central assumption of accelerated life testing.

Related references. Davies (1963, Chapter 8) and Snedecor and Cochran (1967, §§12.5, 12.6) describe methods for the analysis of 3^3 factorial experiments with quantitative factors. Box and Cox (1964) used this example to illustrate the formal estimation of a transformation by maximum-likelihood methods.

Example K

Factorial experiment on diets for chickens

Description of data. An experiment comparing 12 methods of feeding chickens (Duckworth and Carpenter; see John and Quenouille, 1977) was done independently in two replicates arranged in different houses. The treatments, forming a $3 \times 2 \times 2$ factorial, were 'form of protein', 'level of protein', 'level of fish solubles'. The data are given in Table K.1.

Table K.1. Total weights of 16 six-week-old chicks (g)

Protein	Level of protein	Level of fish solubles	House I	House II
Groundnut	0	0	6559	6292
		1	7075	6779
	1	0	6564	6622
		1	7528	6856
	2	0	6738	6444
		1	7333	6361
Soyabean	0	0	7094	7053
		1	8005	7657
	1	0	6943	6249
		1	7359	7292
	2	0	6748	6422
		1	6764	6560

General considerations. Because of the balanced nature of the data the conclusions follow directly from marginal, two-way, etc., tables of mean values. Calculation and inspection of these is an essential first step.

Analysis of variance in such situations serves two purposes. One is to determine an estimate of variance for assessing the precision of the contrasts of means. The second is to ensure that no contrast estimable from the design is overlooked. Usually in factorial experiments it is hoped that main effects and perhaps some two-factor interactions will turn out to be the only appreciable contrasts. However, careful inspection of the full analysis-of-variance table is an advisable precaution against unanticipated features, such as that the data

103

Applied statistics

are best split into separate sections by the level of one of the factors, or that one particular combination of factor levels gives a response quite different from the remaining combinations.

In the present case, three of the factors represent treatments and one, 'houses', replication of the experiment and choice of an estimate of error is most reasonably based on the interactions with houses. This essentially hinges on the supposition that the treatment effects are the same in the two houses; even so it will be wise to check that the various component interactions with houses are roughly comparable.

The analysis. Two-way means are given in Table K.2. The marginal pattern in the two houses is very similar, high weight being achieved for soyabean, low

Table K.2. Means of total weights of 16 chicks (g)

		House	
		I	II
Protein	Groundnut	6966	6559
	Soyabean	7152	6872
Level of	0	7183	6945
protein	1	7099	6755
	2	6896	6447
Level of fish	0	6774	6514
solubles	1	7344	6918

Protein	Level of protein			
	0	1	2	Mean
Groundnut	6676	6892	6719	6763
Soyabean	7452	6961	6624	7012
Mean	7064	6927	6671	6887

Protein	Level of fish solubles		
	0	1	Mean
Groundnut	6537	6989	6763
Soyabean	6751	7273	7012
Mean	6644	7131	6887

Level of fish solubles	Level of protein			
	0	1	2	Mean
0	6750	6594	6588	6644
1	7379	7259	6755	7131
Mean	7064	6927	6671	6887

Table K.3(a). Mean squares for factorial effects within houses, pooled over houses and interactions with houses

	d.f.	m.s. ($\times 10^{-6}$)			
		House I	House II	Main component	Interaction with houses
P	1	0.1038	0.2942	0.3738	0.0242
L_P	2	0.0873	0.2531	0.3181	0.0223
L_F	1	0.9736	0.4892	1.4216	0.0412
$P \times L_P$	2	0.2610	0.1931	0.4291	0.0250
$L_P \times L_F$	2	0.0525	0.1084	0.1544	0.0642
$P \times L_F$	1	0.0447	0.1096	0.0072	0.1471
$P \times L_P \times L_F$	2	0.0766	0.0348	0.0251	0.0864

level of protein and high level of fish solubles. There is a suggestion of interaction between protein and level of protein.

Table K.3 (a) shows the mean squares for the factorial contrasts, first calculated separately within each house and then, more meaningfully, split into a 'main component' based on a pooling of the two houses and an interaction with houses. The interactions with houses (last column) are broadly comparable and provide a pooled estimate of error for the final analysis of variance in Table K.3(b).

The interaction protein × level of protein is significant at the 1 per cent level. Also, level of protein × level of fish solubles is suggestive at the 5 per cent level. We cannot therefore interpret the main effects in a straightforward way. Partitioning the contrasts associated with level of protein into linear and quadratic components is in this instance unenlightening.

Table K.3(b). Mean squares for factorial effects with pooled error

	d.f.	m.s. ($\times 10^{-6}$)
P	1	0.3738
L_P	2	0.3181
L_F	1	1.4216
H	1	0.7083
$P \times L_P$	2	0.4291
$L_P \times L_F$	2	0.1544
$P \times L_F$	1	0.0072
$P \times L_P \times L_F$	2	0.0251
Error	11	0.0448

P, protein; L_P, level of protein; L_F, level of fish solubles; H, house

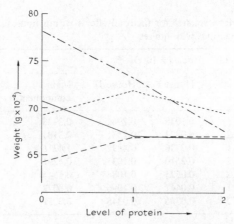

Fig. K.1. Weight of 16 six-week-old chicks versus level of protein.
— — — — —   Groundnut; Level of fish solubles, 0
. . . .   Groundnut; Level of fish solubles, 1
————————   Soyabean; Level of fish solubles, 0
—.—.—.—.   Soyabean; Level of fish solubles, 1

Figure K.1 shows how the weight, averaged over the houses, varies with the $3 \times 2 \times 2$ different diets. The diet producing the highest mean weight is: Protein, soyabean; Level of protein, 0; Level of fish solubles, 1. The average total weight of 16 six-week-old chicks with this diet is 7831 g; the next highest average is 7326 g. The estimated standard error of the difference between these means is 212 g with 11 degrees of freedom.

Further points and exercises.
(i) In the general considerations above, two possibilities are mentioned where description in terms of main effects, etc., is not appropriate for the final presentation of conclusions. How could these possibilities be detected and analysed?

Related reference. John and Quenouille (1977, §3.6) give an analysis of this example.

Example L

Binary preference data for detergent use

Description of data. Table L.1 (Ries and Smith, 1963) compares two detergents, a new product X and a standard product M. Each individual expresses a preference between X and M. In the table, Y_j is the number of individuals out of n_j in 'cell' j who prefer X, the remaining $n_j - Y_j$ preferring M. The individuals are classified by three factors, water softness at three levels, temperature at two levels, and a factor whose two levels correspond to previous experience and no previous experience with M. The object is to study how preferences for X vary.

Table L.1. Number Y_j of preferences for brand X out of n_j individuals

Water softness		M previous non-user		M previous user	
		Temperature		Temperature	
		Low	High	Low	High
Hard	Y_j	68	42	37	24
	n_j	110	72	89	67
Medium	Y_j	66	33	47	23
	n_j	116	56	102	70
Soft	Y_j	63	29	57	19
	n_j	116	56	106	48

General considerations. Here preference for brand X is a binary response variable for each individual and the other variables are explanatory variables. Thus the data are in the form of a $3 \times 2 \times 2$ factorial system, with a considerable number of individuals in each cell, although the unequal numbers in the different cells imply a lack of balance. The three levels of water softness are ranked and some account must be taken of this, even if only informally. A simple, somewhat arbitrary but nevertheless effective device is to treat the three levels as equally spaced on some notional scale of softness and to extract linear and quadratic components of regression.

It is common to analyse binary data like this by a linear logistic model, i.e. by concentrating not so much on θ, the probability of preferring brand X in a particular cell, as on the logistic transform $\log\{\theta/(1-\theta)\}$. There are reasonably cogent arguments for doing this in general, most notably that the parameters so involved are more likely to have a stable interpretation over a range of overall probability levels. The function $\log\{\theta/(1-\theta)\}$ is, however, very nearly linear over the range of θ from 0.2 to 0.8 and in the present data the cell proportions of preferences for brand X are well within that range. Thus a linear model for $\log\{\theta/(1-\theta)\}$ and a linear model for θ are mathematically virtually identical. The analysis directly in terms of θ has the advantage of more direct interpretability and is thus to be preferred in this case. If, however, it were likely that at some stage the data were to be compared with another similar set of data with much higher or much lower preference rates, then estimation on a logistic scale (i.e. effectively multiplication of estimates on a linear scale by an appropriate factor) would make sense.

The proportions of preferences in the different cells are not of equal precision. For fitting a saturated model, i.e. one with as many parameters as cells, these changes of variance are unimportant. For fitting reduced models, i.e. for estimating certain contrasts assuming others to be zero, revised estimates would be obtained by allowing for the changes of variance, fitting by maximum likelihood or weighted least squares. In the present case, however, the changes in variance are relatively minor and it is unlikely that any change of importance would be made by introducing the extra complication. The simpler methods based on unweighted averages of proportions have thus been preferred. Note particularly that in estimating main effects in the analysis below, proportions in the various 'cells' are combined by unweighted averaging.

Methodological details. If in a particular cell there are Y_j preferences for brand X in n_j individuals, the relevant probability is estimated by $p_j = Y_j/n_j$ with an unbiased estimate of variance $v_j = \{Y_j(n_j - Y_j)\}/\{n_j^2(n_j-1)\}$. In the following analysis the average of the v_j's is used as an effective error mean square. The

Table L.2. Observed proportions and estimated variances

Water	M previous non-user		M previous user	
	Temperature		Temperature	
	Low	High	Low	High
Hard	0.618 2	0.583 3	0.415 7	0.358 2
	0.002 17	0.003 42	0.002 76	0.003 48
Medium	0.569 0	0.589 3	0.460 8	0.328 6
	0.002 13	0.004 40	0.002 46	0.003 20
Soft	0.543 1	0.517 9	0.537 7	0.395 8
	0.002 16	0.004 54	0.002 37	0.005 09

Example L 109

main systematic difference in the variances is that the variances are greater at the high temperature than at the low temperature because of the smaller numbers of individuals.

The estimate of variance is based on the binomial distribution, i.e. on the assumed independence of individuals. Any form of 'clustering' in sampling the individuals would tend to inflate variance. If it had been the case that some individuals appeared in several cells, the effective variance would be decreased; so far as we know, all individuals are different.

Table L.3. Two-way and marginal mean

Water	Temperature			Water	M		
	Low	High			Previous non-user	Previous user	
Hard	0.517	0.471	0.494	Hard	0.601	0.387	0.494
Medium	0.515	0.459	0.487	Medium	0.579	0.395	0.487
Soft	0.540	0.457	0.499	Soft	0.530	0.467	0.499
	0.524	0.462	0.493		0.570	0.416	

Temp.	M		
	Previous non-user	Previous user	
Low	0.577	0.471	0.524
High	0.564	0.361	0.462
	0.570	0.416	

The analysis. Table L.2 shows the proportions of preference for brand X, cell by cell, and the associated variances. The average variance is 0.003 181. Table L.3 gives the two-way and marginal proportions obtained by unweighted averaging of the entries of Table L.2.

The main descriptive conclusions stand out from Table L.3. The previous non-users of M have higher proportions preferring X than the previous users of M. Moreover, the changes in proportion for the previous non-users are quite small and different from those for the previous users. For the latter the proportions preferring X are nearly all below $\frac{1}{2}$, are lower at the higher temperature and lower for hard water than for soft water.

To investigate the precision with which these effects are established, the standard factorial contrasts have been calculated from the proportions. Standard errors could be attached to these. Alternatively, Table L.4 gives the analysis-of-variance decomposition into single degrees of freedom.

The main effect of previous usage of M is overwhelmingly significant, as to be expected. Of the other effects, temperature and two interactions with previous

Table L.4. Analyses of variance
(a) Full data

	d.f.	m.s.
Previous usage of M (M)	1	0.071 15
Temperature (T)	1	0.011 49
Softness (linear) (S_L)	1	0.000 05
Softness (quad.) (S_Q)	1	0.000 23
$M \times T$	1	0.007 10
$M \times S_L$	1	0.011 26
$M \times S_Q$	1	0.001 39
$T \times S_L$	1	0.000 70
$T \times S_Q$	1	0.000 05
$M \times T \times S_L$	1	0.001 11
$M \times T \times S_Q$	1	0.001 14
Total	11	0.009 61
Theoretical error	∞	0.003 18

(b) Split data

	d.f.	Previous non-users of M m.s.	Previous users of M m.s.
T	1	0.000 26	0.018 33
S_L	1	0.004 94	0.006 37
S_Q	1	0.000 24	0.001 38
$T \times S_L$	1	0.000 02	0.001 78
$T \times S_Q$	1	0.000 84	0.000 35
Total	5	0.001 26	0.005 64
Theoretical error	∞	0.003 14	0.003 23

usage of M are appreciably greater than error but short of significance at the 5 per cent level. Thus interpretation beyond that of M has to be made with reservations.

If, however, in view of the large main effect of previous use of M, and of two suggestive interactions with it, we split the data into two halves to be analysed separately, the further analyses of Table L.4 are obtained. For the previous non-users the variations are rather less than would be expected by chance. For the previous users, the effect of temperature is now significant at the 2 per cent level, but in judging the strength of evidence for this effect some allowance for selection is desirable.

Example L   111

To summarize, there is clear difference in overall preference proportion for X as between previous non-users of M and previous users of M. Other effects are not definitely established but for the previous users there is a higher proportion at the lower temperature, 0.471 versus 0.361, and a tendency for the proportion to be higher with soft water.

All conclusions about precision depend on the validity of the independence assumptions underlying the calculation of error; however, the fact that the changes in proportion, except for the effect of previous usage, are relatively small may mean that the effects, even if real, are unimportant.

Further points and exercises

(i) Calculate the constant necessary to convert effects on a direct proportion scale into those on a logistic scale, in the range near 50 per cent response.

(ii) Compare the analysis with those given by Bishop, Fienberg and Holland (1975, pp. 142–167); Cox (1970, p. 39); and by Ries and Smith (1963).

Example M

Fertilizer experiment on growth of cauliflowers

Description of data. In an experiment on the effect of nitrogen and potassium upon the growth of cauliflowers, four levels of nitrogen and two levels of potassium were tested:

Nitrogen levels: 0, 60, 120, 180 units per acre (coded as 0, 1, 2, 3);
Potassium levels: 200, 300 units per acre (coded as A, B).

The experiment was arranged in 4 blocks, each containing 4 plots as shown below. When harvested, the cauliflowers were graded according to size. Table M.1 shows the yield (number of cauliflowers) of different sizes: grade 12, for example, means that 12 cauliflowers fit into a standard size crate. The data were provided by Mr J.C. Gower, Rothamsted Experimental Station.

Table M.1. Numbers of cauliflowers of each grade

Block	Treatment	Grade				Unmarketable
		12	16	24	30	
I	0A	—	1	21	24	2
	2B	1	6	24	13	4
	1B	—	4	28	12	4
	3A	1	10	26	9	1
II	3B	—	4	26	14	4
	1A	—	5	27	13	3
	0B	—	—	12	28	8
	2A	—	5	35	5	3
III	1B	—	1	22	22	3
	0A	—	1	8	33	3
	3A	—	6	22	17	2
	2B	—	3	27	14	4
IV	0B	—	—	8	30	10
	2A	—	7	16	22	3
	3B	—	2	31	11	4
	1A	—	—	13	26	9

General considerations. The response variable here is a frequency distribution of cauliflowers of various sizes. After preliminary inspection of the data, the first step is to form one or more derived response variables in terms of which treatments can be compared.

A technical point in the analysis concerns the design used and the estimation of contrasts and error. The treatments form a 4×2 system and one might therefore expect the experiment to be laid out with eight units per block. Because there are in fact only four units per block, not all contrasts can be estimated simply from within-block comparisons and some close study of the design employed is therefore needed.

The analysis. The frequency distributions in Table M.1 show systematic differences between treatments; level 3 of nitrogen, for example, produces more high-grade cauliflowers than does level 0. To compare the treatments, we may form derived response variables in various ways. Two such derived variables are given in Equations (M.1) and (M.2).

For each plot we may, for instance, calculate

$$Y = \frac{1}{12}n_{12} + \frac{1}{16}n_{16} + \frac{1}{24}n_{24} + \frac{1}{30}n_{30}, \tag{M.1}$$

where n_r ($r = 12, 16, 24, 30$) denotes the observed frequency of cauliflowers of grade r. Then Y is the effective number of crates of marketable cauliflowers. It could be modified by introducing market values as weighting factors. Alternatively, we can take

$$Z = (n_{12} + n_{16} + n_{24})/48, \tag{M.2}$$

this being the proportion of cauliflowers of grade 24 or better; the divisor 48 corresponds to the total number of cauliflowers per plot except in those instances in which some plants died.

Here we consider the analysis only of response variable (M.1). An analysis based upon Equation (M.2) using a logistic model is likely to lead to similar conclusions.

Table M.2. Yield of cauliflowers, derived response Y

Treatment	Block		Treatment	Block	
	I	III		II	IV
0A	1.738	1.496	0B	1.433	1.333
1B	1.817	1.712	1A	1.871	1.408
2B	1.892	1.779	2A	1.938	1.838
3A	2.092	1.858	3B	1.800	1.783

The values of Y for the sixteen plots are shown in Table M.2.

The experimental design is a complete 2×4 factorial, confounded into two blocks (I, II, say) with the other two blocks (III, IV) being a replicate. In order to determine the system of confounding we look at the coefficients of linear contrasts for the 2×4 factorial. These are given in Table M.3 and the $K \times N_Q$ interaction is seen to define the confounding.

Table M.3. Contrasts for 2×4 factorial

Potassium level	A				B			
Nitrogen level	0	1	2	3	0	1	2	3
Main effects:								
Potassium (K)	-1	-1	-1	-1	1	1	1	1
Nitrogen, Linear (N_L)	-3	-1	1	3	-3	-1	1	3
Quadratic (N_Q)	1	-1	-1	1	1	-1	-1	1
Cubic (N_C)	-1	3	-3	1	-1	3	-3	1
Interactions:								
$K \times N_L$	3	1	-1	-3	-3	-1	1	3
$K \times N_Q$	-1	1	1	-1	1	-1	-1	1
$K \times N_C$	1	-3	3	-1	-1	3	-3	1

In principle, an estimate of error for any treatment contrast is provided by its interaction with replicates and, under the assumption of homogeneity, all such interactions can be pooled, giving an error mean square with 6 degrees of freedom, i.e. interactions with all contrasts in Table M.3 except $K \times N_Q$ ($\equiv$ blocks). Also we may pool the cubic contrast N_C and its interaction with replicates, thus giving an error with 8 degrees of freedom. This leads to the analysis of variance given in Table M.4.

The nitrogen linear contrast N_L is significant at the 0.1 per cent level. No other contrast is statistically significant, although the mean squares for N_Q and K are each greater than the error mean square by a factor of about three. Estimates of any unconfounded contrasts can be obtained directly from the mean values of the observations in the usual way.

Table M.4. Analysis of variance, derived response Y

	d.f.	s.s.	m.s.
Blocks	3	0.1774	0.0591
Main effects: K	1	0.0295	
N_L	1	0.3429	
N_Q	1	0.0325	
Interaction: $K \times N_L$	1	0.0000	
Error	8	0.0866	0.0108
Total	15	0.6689	

An analysis of residuals gives no evidence of outliers or nonnormality.

A summary of the conclusions from the analysis is as follows:

(i) There is no evidence that increasing the level of potassium from 200 to 300 units per acre increases the yield of cauliflowers, as measured by the derived variable Y. In fact, the mean is 0.086 crates lower (est. s.e. = 0.052 with 8 d.f.) at the higher potassium level.

(ii) Increasing the level of nitrogen increases the yield of cauliflowers. The estimated average yield, assuming a quadratic trend over the experimental levels, is given in Table M.5. Increasing nitrogen from 60 to 120 units per acre increases the yield significantly by $1.85 - 1.72 = 0.13$ crate (est. s.e. = 0.024 with 8 d.f.). No further significant improvement is achieved by increasing nitrogen to 180 units per acre.

Table M.5. Estimated yield, derived response Y

Nitrogen level (units per acre)	0	60	120	180
Estimated yield (no. crates)	1.49	1.72	1.85	1.89

In the above analysis the dependence on nitrogen level has been assumed, for simplicity, to be locally quadratic. It would be possible to fit a more complex functional relationship, such as one representing growth to a limit, which would be more plausible on general grounds. This is likely to be worthwhile, however, only if careful comparison is required of these data with other data obtained over a different range of levels.

Further points and exercises

(i) Analyse the data in terms of the derived response variable (M.2).

(ii) It is suggested that the numbers of cauliflowers of the five grades should be treated as a five-dimensional response variable and the techniques of formal multivariate analysis, in particular canonical regression analysis, used. Criticize this proposal, ignoring special aspects introduced by the confounding.

Related reference. Davies (1963, Chapters 8 and 9) discusses contrasts and the analysis of experiments involving confounding.

Example N

Subjective preference data on soap pads

Description of data. Table N.1 gives data obtained during the development of a soap pad. The factors, amount of detergent, D, coarseness of pad, C, and solubility of detergent, S, were each set at two levels. There were 32 judges and the experiment was done on two days. Each judge attached a score (excellent $= 1, \ldots$, poor $= 5$) to two differently formulated pads on each of two days. For the data and several different analyses, see Johnson (1967).

General considerations. There are two important special features about these data. Firstly, responses are recorded on a qualitatively ordered scale on which, for example, there is no guarantee that the difference between, say, 1 and 2 is meaningfully comparable with the difference between 3 and 4. Secondly, a rather complex (and in many ways inappropriate) experimental design has been used.

If the data were of simple structure it might well be feasible to use a primarily 'distribution-free' approach. In the present instance, however, it is easier to start by an analysis of means, treating the responses as an ordinary quantitative variable. A rough examination of the distribution of responses can be used to supplement the analysis of means.

The design used here involves a particular scheme of confounding. We shall discuss analysis from first principles rather than using the specialized results of the theory of confounded designs. A crucial point concerns the role of differences between judges. If these are relatively minor, we may regard the mean score given by a judge to a particular treatment as a derived reponse; because each judge looks only at two treatments, differences between judges are rather poorly determined. A more cautious approach, however, is to eliminate differences between judges by taking as derived response variable the difference in means between the two treatments for a particular judge.

The analysis. Table N.2 gives combined means for the four replicates, day 1 and day 2; there are no clear patterns or major differences and for much of the remainder of the analysis we do not specifically distinguish replicates and days. Table N.3 shows the frequency distribution of the individual responses for the 8 treatments. The 16 values for each treatments are from 8 judges, each measuring twice.

116

Table N.1. Subjective scores allocated to soap pads prepared in accordance with 2^3 factorial scheme. Five-point scale; 1 = excellent, 5 = poor

Judge	Treatment	Day 1	Day 2	Judge	Treatment	Day 1	Day 2
Replicate I				Replicate II			
1	1	2	4	5	1	4	2
17	1	2	3	21	1	3	3
1	*dcs*	4	4	5	*cs*	3	4
17	*dcs*	4	4	21	*cs*	1	2
2	*d*	5	4	6	*d*	1	2
18	*d*	4	4	22	*d*	5	4
2	*cs*	2	1	6	*dcs*	3	3
18	*cs*	1	2	22	*dcs*	4	4
3	*c*	1	3	7	*c*	3	3
19	*c*	5	5	23	*c*	3	5
3	*ds*	3	2	7	*s*	4	4
19	*ds*	4	3	23	*s*	5	3
4	*s*	1	3	8	*dc*	4	4
20	*s*	2	3	24	*dc*	2	3
4	*dc*	3	4	8	*ds*	3	2
20	*dc*	3	3	24	*ds*	2	3
Replicate III				Replicate IV			
9	1	3	2	13	1	3	4
25	1	2	3	29	1	3	4
9	*ds*	4	3	13	*dc*	2	3
25	*ds*	3	3	29	*dc*	3	4
10	*d*	1	1	14	*d*	4	4
26	*d*	3	3	30	*d*	4	3
10	*s*	2	1	14	*c*	2	2
26	*s*	1	1	30	*c*	4	5
11	*c*	3	3	15	*s*	5	5
27	*c*	3	3	31	*s*	3	3
11	*dcs*	3	3	15	*dcs*	4	4
27	*dcs*	2	2	31	*dcs*	1	2
12	*dc*	3	3	16	*ds*	4	3
28	*dc*	4	4	32	*ds*	4	3
12	*cs*	1	2	16	*cs*	3	4
28	*cs*	3	3	32	*cs*	1	4

Table N.2. Means for replicate–day combinations

Replicate	1	2	3	4
Day 1	2.875	3.125	2.562	3.125
Day 2	3.250	3.188	2.500	3.562

From Table N.3 it is apparent that there are no striking differences between treatments. In terms of mean response, the combination *sc* does best with little to choose between the other combinations. If we concentrate on the scores 1, and possibly 2, which in practice may well be the most interesting ones, treatments *s* and *sc* seem the best.

Inspection of the results suggests there are no major differences between judges, except for judges 10 and 26 in replicate 3 who have given treatments *d* and *s* very favourable scores. Without specific information about the nature of the judges, there is no reason to 'reject' these values. It is reasonable in an approximate analysis to interpret the mean values by an analysis of variance as if there were 16 separate judges in each replicate rather than 8 judges, each acting for two treatments. The net effect is to favour somewhat the combination

Table N.3. Frequency distributions of response versus treatment

	Treatment							
	1	s	c	sc	d	ds	dc	dsc
Score								
1	0	4	1	5	3	0	0	1
2	5	2	2	4	1	3	2	3
3	7	5	8	4	3	9	8	4
4	4	2	1	3	7	4	6	8
5	0	3	4	0	2	0	0	0
Mean	2.938	2.875	3.312	2.312	3.250	3.062	3.250	3.188

d and *s* and to overestimate error. Table N.4 gives the analysis of variance. In the absence of identification of specific qualitative distinctions between replicates, treatments × replicates (m.s. = 1.811) provides a valid cautious estimate of the error of treatment contrasts. It is clear without formal testing that the treatment differences (m.s. = 1.713) are entirely explicable as random variation. The split into factorial contrasts is not particularly helpful. Indeed, the relatively large value for the three-factor interaction is a warning of the inappropriateness of the factorial representation: the simplest descriptive comment on the treatment means is that the combination *cs* has a relatively good score with little variation between the other 7 treatments.

The last two lines of the analysis of variance throw some light on the error structure. If σ^2 is the variance of pure error for one judge and σ_J^2 the variance of systematic judge effects, we have the estimates

$$\tilde{\sigma}^2 = 0.367, \qquad \tilde{\sigma}^2 + 2\tilde{\sigma}_J^2 = 1.742,$$
$$\text{i.e. } \tilde{\sigma} = 0.606, \qquad \tilde{\sigma}_J = 0.829.$$

The meaningfulness of these estimates depends strongly on the precautions of randomization and concealment used to ensure independence.

The relatively large value of $\tilde{\sigma}_J$ implies that it is probably worth making an analysis in which judge effects are eliminated, the previous analysis being 'valid' but inefficient. In view of the remarks about judges 10 and 26, one of the main changes brought about by using a within-judge analysis is likely to be an increase in the estimated means for d and for s.

Table N.4. Analysis of variance ignoring repetition of judges

	d.f.	s.s.	m.s.
S	1	3.4453	
C	1	0.0078	
D	1	3.4453	
$C \times D$	1	0.1953	
$D \times S$	1	1.3203	
$S \times C$	1	1.3203	
$S \times C \times D$	1	2.2578	
Treatments	7	11.9922	1.713
Replicates	3	11.6484	3.883
Treatments × Replicates	21	38.0394	1.811
Days	1	1.3203	1.320
Days × Replicates	3	1.3983	0.466
Days × Treatments	7	6.7422	0.963
Days × Treatments × Replicates	21	10.2891	0.490
Between judges, within treatments, within replicates	32	55.7500	1.742
Days × Between judges	32	11.7500	0.367
Total	127		

For a relatively simple within-judge analysis, a derived response is taken giving the difference in total response for a particular judge. Thus, for judge 1, $4+4-2-4$ is taken as a response estimating θ_{dcs}, the parameter comparing *dcs* with 1. Similarly for judge 2, $2+1-5-4$ estimates $\theta_{cs} - \theta_d$, etc. Note that the θ's have to be halved to give differences of mean response, and test under the simplest notions about error will have variance $4\sigma^2$ estimated as 1.468. Table N.5 summarizes the results of the least-squares analysis. The F value for fitting parameters is 2.19, which is slightly larger than the 10 per cent point. There is thus a sharpening of the apparent precision by the use of the within-judge analysis, but apparent differences still cannot be claimed as unambiguously

Table N.5. Within-judge analysis of derived responses

	d.f.	s.s.	m.s.
Fitting parameters	7	78.83	11.26
Residual	25	128.17	5.13
Between sessions	9	56.67	6.30
Between pairs within sessions	16	71.50	4.47
Total	32	207.00	

Estimated parameters

1	s	c	sc	d	ds	dc	dcs
—	1.79	1.58	−1.38	2.62	0.42	1.46	1.50

established by the experiment. To interpret the θ's, they have been halved and then adjusted to agree in overall mean with the data, giving the second row of Table N.6. The adjustment has enhanced the apparent superiority of cs and, as anticipated, worsened d and s.

To summarize, the experiment does not clearly establish effects of the three factors studied. The dispersion of response (Table N.3) is considerable. The main difference suggested by inspection of means (Table N.6) is between cs and the other combinations.

Table N.6. Comparison of means from two analyses

	1	s	c	sc	d	ds	dc	dcs
Crude mean	2.94	2.88	3.31	2.31	3.25	3.06	3.25	3.19
Adjusted mean	2.52	3.42	3.31	1.84	3.84	2.73	3.25	3.27

Example O   Atomic weight of iodine

Description of data. Table O.1 gives ratios of reacting weight of iodine and silver obtained, in an accurate determination of atomic weight of iodine, using five batches of silver A, B, C, D, E and two of iodine, I, II (Baxter and Landstredt, 1940; Brownlee, 1965). Silver batch C is a repurification of batch B, which in turn is a repurification of batch A. In these data 1.176 399 has been subtracted from all values.

Table O.1. Ratios of reacting weight with 1.176 399 subtracted $\times 10^6$

Silver batch	Iodine batch	
	I	II
A	23, 26	0, 41, 19
B	42, 42	24, 14
C	30, 21, 38	
D	50, 51	62
E	56	

General considerations. These data illustrate in very simple form some of the issues in analysing unbalanced data such as arise commonly, although by no means exclusively, in observational studies. Most of the following points apply broadly to the analysis of unbalanced data with two or more factors of cross-classification.

The questions to be considered with unbalanced data are essentially the same as with analogous balanced data. Is there interaction? If not, convenient summarization is via estimated row and column contrasts and sometimes even via just, say, the row contrasts when the column effects vanish. Now for balanced data, simple row and column means determine estimated row and column contrasts which are such that:

(i) they are descriptively appealing;

(ii) the estimated difference, say, between two rows is unaffected if it is postulated that column effects are absent;

(iii) the row and column means are least-squares estimates under suitable models;

(iv) sums of squares are simply isolated for testing relevant hypotheses and correspond directly to a decomposition of the data vector into orthogonal components.

A small lack of balance can often be satisfactorily dealt with by some *ad hoc* modification of the procedures for balanced data. In general, however, and certainly with the present data, this is either unsatisfactory or impossible.

One central difficulty is that even under a model with no interaction, i.e. with additive row and column effects, the row and column means do not estimate relevant parameters. For example, in silver batch C, the row mean has no contribution from the second column, iodine batch II, and hence is in general biased as an estimate of a row parameter. In fact, all the properties (i)–(iv) break down and it is necessary to proceed by more explicit fitting of a sequence of models. Of course the formulation of suitable models depends on the context, but in problems where normal-theory linear models may be expected to be appropriate, the following forms will be natural for a two-way classification.

Let Y_{ijk} be the kth observation in row i and column j. If there are n_1 rows and n_2 columns with r_{ij} observations in row i and column j, we have $i = 1, \ldots, n_1$; $j = 1, \ldots, n_2$; $k = 1, \ldots, r_{ij}$. Natural models have Y_{ijk} independently normally distributed with constant variance σ^2 and with

$$
\begin{aligned}
&\text{Model I} && E(Y_{ijk}) = \mu; \text{ homogeneity} \\
&\text{Model II}_1 && E(Y_{ijk}) = \mu + \alpha_i; \text{ pure row effects} \\
&\text{Model II}_2 && E(Y_{ijk}) = \mu + \beta_j; \text{ pure column effects} \quad\quad\text{(O.1)} \\
&\text{Model II}_{12} && E(Y_{ijk}) = \mu + \alpha_i + \beta_j; \text{ additivity (no interaction)} \\
&\text{Model III} && E(Y_{ijk}) = \mu_{ij} = \mu + \alpha_i + \beta_j + \gamma_{ij}; \text{ arbitrary means}.
\end{aligned}
$$

The fitting of these models by least squares is entirely straightforward and based on row and column means, except for Model II$_{12}$, where the solution of the least-squares equations is required, with attention to the redundancy of specification; see below. The residual sum of squares from Model III is the sum of squares within cells. When this is subtracted from the residual sum of squares for Model II$_{12}$, the sum of squares for testing the null hypothesis of no interaction (Model II$_{12}$) results. This is an adaptation to the present problem of the general procedure for testing subhypotheses in multiple regression.

If the data show clear evidence of interaction, estimates and interpretation of main-effect parameters will be relevant only in those rather rare circumstances in which one is for some clear practical reason interested in effects of one factor, say rows, averaged over the levels of the other factor, columns. This is not the case in the present application. Suppose, however, that rows represent treatments and columns represent classification of the individuals, e.g. into male and female. In some narrowly technological situations one may then be concerned with the average treatment effect over a population of individuals with specified proportions of males and females; note that these proportions might not be those occurring in the data, and that the proportions used are relevant.

If, as in the present data, an interpretation can be based on the model without interaction, there is the further complication that estimated row effects depend on whether column effects are present in the model. This is typical of multiple-regression calculations in which the estimate of one regression coefficient depends on what other terms are in the model. The apparent precision of the row effects is higher if column effects are omitted, but in the present simple situation in which both row and column effects are presumably of equal interest, it seems more reasonable to present conclusions primarily from the analysis of the full additive model, Model II_{12}.

Some methodological details. As explained above, the calculations for the present analysis are those of least-squares theory, i.e. multiple regression. There is the technical complication that the models as formulated above are redundant, i.e. overparameterized, so that the usual least-squares equations have infinitely many solutions. There are several ways of resolving this difficulty.

For example, the computer program GLIM adopts a conventional parameterization chosen for computational generality and convenience rather than for statistical interpretability. When such parameterizations are used, it is essential that for the presentation of conclusions readily interpreted forms of estimate are given.

If the analysis is done from first principles there are three broad approaches:

(a) reparameterization without redundant parameters;
(b) imposition of constraints;
(c) fitting of parameters in stages, e.g. rows first, then columns.

If the analysis is done via a general multiple-regression program, method (a) must be used. In both methods (a) and (b) there is a gain in simplicity in choosing the reparameterization or constraints to achieve a reduction in the size of the set of linear equations to be solved, but whether this simplicity is worth special effort depends very much on the size of the problem and the computational resources available.

The analysis. The residual sums of squares obtained after fitting the sequence of models (O.1) are given in Table O.2(a) and in equivalent analysis-of-variance form in Table O.2(b). Note that when fitting model II_{12} from first principles it is necessary to impose constraints upon the parameters, e.g. (i) $\Sigma r_{i.}\alpha_i = \Sigma r_{.j}\beta_j = 0$, (ii) $\Sigma \alpha_i = \Sigma \beta_j = 0$, or (iii) $\alpha_1 = \beta_1 = 0$. Different sets of constraints will lead to the same residual sum of squares and to the same values of

Table O.2(a). Residual sums of squares

Model	s.s.	d.f.
III	1041.7	8
II_{12}	1533.2	10
II_1	1683.1	11
II_2	3782.2	14
I	4255.4	15

Table O.2(b). Analysis of variance

	d.f.	s.s.	
Iodine (ignoring Silver)	1	473.2	149.9 Iodine (adjusted for Silver)
Silver (adjusted for Iodine)	4	2249.0	2572.3 Silver (ignoring Iodine)
Iodine and Silver	5	2722.2	
Iodine × Silver	2	491.5	
Between cells	7	3213.7	
Within cells	8	1041.7	
Total	15	4255.4	

meaningful (estimable) functions of parameters, but to different least-squares estimates for the individual parameters.

There is no evidence of any interaction between iodine and silver batches; the *F* ratio for testing this is

$$\frac{491.5/2}{1041.7/8} = 1.89$$

with (2,8) degrees of freedom. Nor is there evidence of differences between iodine batches. For differences between silver batches, the ratio

$$\frac{2249.0/4}{1041.7/8} = 4.32$$

with (4,8) degrees of freedom is significant at the 5 per cent level.

Estimates of silver batch means $(\hat{\mu} + \hat{\alpha}_i)$ are given in Table O.3(a). The estimates obtained fitting model II_{12} differ little from the observed batch means

Table O.3(a). Estimated silver batch means

	A	B	C	D	E
Model II_{12} (adjusted means)	22.5	30.5	26.0	53.1	52.3
II_1 (unadjusted means)	21.8	30.5	29.7	54.3	56.0

$\bar{Y}_{i..}$, i.e. fitting II_1, as is expected in the absence of any significant iodine effect. Estimated standard errors of batch means, $\tilde{\sigma}/\sqrt{r_i}$. with $\tilde{\sigma}$ equal to $\sqrt{(1041.7/8)}$, are given in Table O.3(b). Silver batches D and E give significantly higher results than do batches A, B and C.

Table O.3(b). Estimated standard errors of batch means (d.f. = 8)

A	B	C	D	E
5.1	5.7	6.6	6.6	11.4

Further points and exercises

(i) Check for fitting model II_{12} that:

(a) imposing constraints $\Sigma r_{i.}\alpha_i = \Sigma r_{.j}\beta_j = 0$, or $\Sigma\alpha_i = \Sigma\beta_j = 0$ or $\alpha_1 = \beta_1 = 0$;

or (b) reparameterizing to avoid redundant parameters;

lead to identical conclusions.

Related references. Armitage (1971, §8.7) and Snedecor and Cochran (1967, §16.7) discuss the analysis of unbalanced two-way tables.

Example P

Multifactor experiment on a nutritive medium

Description of data. Fedorov, Maximov and Bogorov (1968) obtained the data in Table P.1 from an experiment on the composition of a nutritive medium for green sulphur bacteria *chlorobrium thiosulphatophilum*. The bacteria were grown under constant illumination at a temperature of 25–30 °C: the yield was determined during the stationary phase of growth. Each factor was at two levels, with each level used 8 times. Subject to this, the factor levels were randomized.

General considerations. Although these data in fact arose from an experiment, the haphazard character of the design means that the analysis is in many ways more typical of that of unbalanced data arising in observational studies. The availability of an independent external estimate of standard deviation is in the present case very important.

While there is no uniquely optimal way of approaching the analysis, the following is sensible. Fit a model containing only main effects, thus involving 11 parameters. If this produces a residual mean square consistent with the external estimate of variance, interpretation will primarily be via the estimated parameters in that model, supplemented possibly by examination of one or two two-factor interaction terms thought potentially important. If, however, as is actually the case, the main-effect model leads to a residual mean square much too big, some consideration of at least two-factor interactions has to be made. With 10 factors, there are 45 two-factor interactions and with main effects and general mean, there are then 56 parameters, which of course cannot be estimated from 16 observations.

A selection of parameters for fitting must therefore be made. Two general ideas help in the choice:

(a) it will extremely rarely be sensible to introduce a two-factor interaction term without at the same time introducing the corresponding main effects;

(b) it is often the case that two factors both with substantial main effects show appreciable interaction.

The approach is thus to fit a small number of main effects which the first analysis shows to be appreciable, plus the corresponding interaction terms. Various possibilities are tried and in principle all those fairly simple models

Table P.1. Yields of bacteria

	Factors										Y
	x_1 NH$_4$Cl	x_2 KH$_2$PO$_4$	x_3 MgCl$_2$	x_4 NaCl	x_5 CaCl$_2$	x_6 Na$_2$S · 9H$_2$O	x_7 Na$_2$S$_2$O$_3$	x_8 NaHCO$_3$	x_9 FeCl$_3$	x_{10} micro-elements	Yield
Levels $\{$ +	1500	450	900	1500	350	1500	5000	5000	125	15	
−	500	50	100	500	50	500	1000	1000	25	5	
1	−	+	+	+	−	+	−	+	−	+	14.0
2	−	−	+	+	−	+	+	−	−	+	4.0
3	+	−	−	+	+	+	−	−	−	−	7.0
4	−	−	+	−	+	+	−	+	+	+	24.5
5	+	−	+	+	+	+	−	−	−	−	14.5
6	−	−	+	−	+	+	+	+	+	+	71.0
7	+	+	−	+	−	−	−	+	−	−	15.5
8	+	+	−	+	+	−	−	−	+	+	18.0
9	−	+	+	+	−	−	+	+	−	−	17.0
10	+	+	+	+	+	−	+	+	+	+	13.5
11	−	+	−	−	−	−	+	+	−	+	52.0
12	+	+	−	−	+	−	+	−	+	−	48.0
13	+	+	−	−	−	+	+	−	+	−	24.0
14	−	+	−	−	+	−	−	+	+	−	12.0
15	+	−	−	−	−	+	−	+	+	+	13.5
16	−	−	−	+	−	+	+	+	−	+	63.0

All the concentrations are given in mg/l, with the exception of factor 10, whose central level (10 ml of solution of micro-elements per 1 l of medium) corresponds to 10 times the amount of micro-element in Larsen's medium.
The yield has a standard error of 3.8.

giving a residual mean square consistent with the external variance are listed, together with the corresponding parameter estimates.

Of course the possibility remains that no reasonably simple model gives a small enough residual mean square. This would mean that either the situation is a very complicated one or that the external estimate of variability is too small. In fact, it is often the case in practice that nominal standard deviations based perhaps on replicate measurements over a short period omit relevant sources of variability and thus are too small.

The analysis. We code the high and low level of each component constituent $x_1, \ldots, x_{10}$ as $+1$ and -1, respectively, and fit a sequence of multiple regressions to the observations on yield Y, starting with a straightforward

Table P.2. Residual sums of squares for sequence of regression models

Terms included	Residual sum of squares	Degrees of freedom
$x_1, x_2, x_3, x_4, x_5, x_6, x_7, x_8, x_9, x_{10}$	1528.5	5
x_8	4030.7	14
x_7, x_8	2105.7	13
x_7, x_8, x_7x_8	652.2	12
x_7, x_8, x_7x_8, x_9	536.5	11
$x_7, x_8, x_7x_8, x_9, x_6, x_6x_8$	171.9	9

main-effects model. The results are summarized in Table P.2. The main-effects model containing 10 components gives a residual mean square of $1528.5/5 = 305.7$, with 5 degrees of freedom, which is very much higher than the external estimate of variance $(3.8)^2 = 14.44$.

The two components which individually give the best fit to the data are x_8 and x_7, and are in fact orthogonal as can be seen from Table P.1. A regression on x_7, x_8 and the interaction (cross-product) term x_7x_8 gives a reduced residual mean square of $652.2/12 = 54.4$ (12 d.f.), a considerable improvement over the main-effects model, but still an inadequate fit. We next include x_9, the most significant of the omitted components. This gives a residual mean square of $536.5/11 = 48.8$ (11 d.f.). Although x_9 itself is significant, its interactions with x_7 and x_8 are not so. None of the remaining components if introduced as a main effect produces any further significant improvement. If, however, we include also interactions with others already in the model, the addition of x_6 and x_6x_8 reduces the residual mean square to $171.9/9 = 19.1$ (9 d.f.) which, if judged against the external estimate of variance, by comparing the ratio $171.9/14.44 = 11.9$ with the tabulated χ^2-distribution with 9 degrees of freedom, suggests that the fit is reasonable.

The greatest reduction due to a further component and one of its interactions is given by x_3 and x_3x_6, reducing the residual sum of squares to 78.9 (7 d.f.) but, if these are included, the estimated coefficient of x_3 is negative and is not consistent with an increased level improving the nutritive medium.

Thus we adopt the model

$$E(Y) = \beta_0 + \beta_6 x_6 + \beta_7 x_7 + \beta_8 x_8 + \beta_9 x_9 + \beta_{68} x_6 x_8 + \beta_{78} x_7 x_8. \qquad \text{(P.1)}$$

The least-squares estimates and standard errors of the coefficients in Equation (P.1) are given in Table P.3. The covariances of the estimates are

$$\text{cov}(\hat{\beta}_6, \hat{\beta}_7) = -\text{cov}(\hat{\beta}_6, \hat{\beta}_8) = -\text{cov}(\hat{\beta}_7, \hat{\beta}_8) = 0.0694,$$
$$\text{cov}(\hat{\beta}_6, \hat{\beta}_9) = \text{cov}(\hat{\beta}_7, \hat{\beta}_9) = -\text{cov}(\hat{\beta}_8, \hat{\beta}_9) = 0.2777,$$

Table P.3. Estimates and standard errors for final regression model

Coefficient	Estimate	Standard error
β_0	25.72	0.95
β_6	1.35	0.99
β_7	11.78	0.99
β_8	11.47	0.99
β_9	3.26	1.05
β_{68}	4.59	0.95
β_{78}	9.53	0.95

other covariances being negligible. Fitted values based on Equation (P.1) and residuals are given in Table P.4. Figure P.1 shows the residuals plotted against expected normal order statistics; the curvature towards the extremes possibly reflects the effect of fitting many parameters to a small amount of data.

Table P.4. Observed and fitted values

	Observed	Fitted	Residual		Observed	Fitted	Residual
1	14.0	18.6	-4.6	9	17.0	16.5	0.5
2	4.0	10.0	-6.0	10	13.5	18.5	-5.0
3	7.0	5.5	1.5	11	52.0	55.8	-3.8
4	24.5	25.1	-0.6	12	48.0	49.3	-1.3
5	14.5	10.0	4.5	13	24.0	23.0	1.0
6	71.0	67.7	3.3	14	12.0	12.0	0.0
7	15.5	12.0	3.5	15	13.5	13.2	0.3
8	18.0	13.2	4.8	16	63.0	61.2	1.8

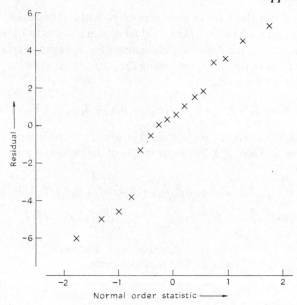

Fig. P.1. Residuals of yield from model (P.1) versus normal order statistics.

The maximum predicted yield under Equation (P.1), assuming two levels for each component, is obtained with each of x_6, x_7, x_8 and x_9 at its high level, the predicted yield then being 67.7 with standard error 2.6. The sixth row of Table P.1 corresponds to an experimental point under these conditions and the observed yield is 71.0 in reasonably close agreement with the fitted value.

Further points and exercises

(i) What design would be more suitable for investigating 10 two-level factors in 16 observations?

(ii) Consider the computational problem of searching more systematically for all simple models consistent with the external estimate of variability.

(iii) How might the analysis have proceeded had the external estimate of variability not been available?

(iv) Might it have been better to study the dependence of log yield on log concentrations?

Example Q   **Strength of cotton yarn**

Description of data. An experiment was done with the objects of estimating:
(i) the difference in mean strength of two worsted yarns produced by slightly
different processes, and (ii) the variation of strength between and within
bobbins for yarns of this type. For each yarn a considerable number of bobbins were produced and 6 bobbins selected at random. From each of these,
4 short lengths were chosen at random for strength testing. The breaking
loads are given in Table Q.1.

Table Q.1. Breaking loads (oz)

Bobbin	1	2	3	4	5	6
Yarn A	15.0	15.7	14.8	14.9	13.0	15.9
	17.0	15.6	15.8	14.2	16.2	15.6
	13.8	17.6	18.2	15.0	16.4	15.0
	15.5	17.1	16.0	12.8	14.8	15.5
Yarn B	18.2	17.2	15.2	15.6	19.2	16.2
	16.8	18.5	15.9	16.0	18.0	15.9
	18.1	15.0	14.5	15.2	17.0	14.9
	17.0	16.2	14.2	14.9	16.9	15.5

General considerations. This problem illustrates in fairly simple form the
analysis of data in which the pattern of random variation has some nontrivial
structure. Because of the nature of the response variable, it is reasonable to
concentrate on means and variances, this being totally appropriate if the
variation is normal. It would be possible to use broadly similar, although more
complicated, techniques based on detailed models involving particular non-
normal distributions, Poisson, binomial, exponential, etc., were that appropriate.

The numbering of bobbins within each yarn is random, as too is the numbering of test lengths within bobbins. Thus the analysis-of-variance table has the
doubly nested form shown in Table Q.4. Note that the form of the table is
settled by the way the data were obtained. It would be possible to examine

differences between individual bobbins, although this would normally be fruitful only if more information were available to characterize the bobbins as individuals. Because the bobbins are chosen from a large population, it is helpful to regard between-bobbin variation as a form of random variation, described by a component of variance between bobbins, i.e. the variance over the population of the 'true' bobbin means.

That is, we characterize the random variation for each yarn by two components of variance, one between lengths within bobbins and the other between bobbins. These are estimated from the analysis-of-variance table, either separately for each yarn or pooled over yarns.

More broadly, there are four possibilities that could have arisen:

(i) the bobbins are identifiable individuals;
(ii) only the variance of the 'true' means, calculated over the bobbins individually observed, is of interest;
(iii) the bobbins are a sample from a finite population of known size and only the variance of the 'true' means, calculated over the finite population, is of interest,
(iv) the population of bobbins is effectively infinite and again the variance of the 'true' means is required.

The analysis-of-variance table is the same in all cases; it is the interpretation of the bobbin means that is different. More than one of (i)–(iv) might be relevant for different purposes, although in fact (iv) is the most useful here.

The components of variance have at least three uses. Firstly, they provide summary descriptions of important aspects of the problem under study and so may be regarded as primary parameters. Next, they clarify the fact that to compare the means of the two yarns, the mean square between bobbins within yarns provides an appropriate estimate of error. Finally, via a process of synthesis of variance, it is possible to estimate the consequences of some different scheme of sampling in which the number of repeat observations per bobbin is changed.

To see that the analysis of variance does not overlook some major aspect of importance, it is desirable to check that:

(a) no single observation or small number of 'wild' observations has inflated the mean square within bobbins;
(b) no single bobbin has inflated the mean square between bobbins;
(c) there is no relation between the bobbin means and standard deviations that might suggest a transformation;
(d) the between-bobbin variation is approximately normal.

It is possible also to examine for each yarn the variation from bobbin to bobbin in the mean square within bobbins. Because the bobbins are randomly numbered, excessive variation in these mean squares would imply either some non-independence in sampling or a long-tailed distribution within bobbins.

In principle it would be possible to make similar analyses of higher cumulants.

The analysis. Mean breaking loads for each bobbin and yarn are summarized in Table Q.2. The overall mean for yarn B is 0.86 oz higher than that for yarn A.

Table Q.2. Mean breaking loads (oz)

Bobbin		1	2	3	4	5	6	Mean
Yarn	A	15.32	16.50	16.20	14.22	15.10	15.50	15.47
	B	17.52	16.72	14.95	15.42	17.78	15.62	16.33

Table Q.3 gives the mean squares within bobbins. There is no evidence of any systematic relationship between these and the bobbin means, nor is there evidence of heterogeneity between the mean squares within either yarn. Approximate normality of the variation between bobbins is confirmed by plotting bobbin means against expected normal order statistics (Pearson and Hartley, 1966, Table 28).

Table Q.3. Mean squares within bobbins (3 d.f.)

Bobbin		1	2	3	4	5	6
Yarn	A	1.76	1.01	2.05	1.03	2.47	0.14
	B	0.53	2.21	0.58	0.23	1.15	0.32

The analysis of variance is given in Table Q.4. There is no significant difference in variation according to yarn, either within bobbins ($F = 1.41/0.83 = 1.69$ with (18, 18) d.f.) or between bobbins ($F = 5.51/2.64 = 2.09$ with (5, 5) d.f.). It is convenient to pool mean squares within yarns.

Table Q.4. Analysis of variance

	d.f.	s.s.	m.s.	E(m.s.)
Between yarns	1	8.927		
Between bobbins: within yarn A	5	13.210	2.64	
within yarn B	5	27.569	5.51	
Between bobbins within yarns	10	40.779	4.08	$\sigma_0^2 + 4\sigma_1^2$
Within bobbins: within yarn A	18	25.355	1.41	
within yarn B	18	15.027	0.83	
Within bobbins	36	40.382	1.12	σ_0^2
Total	47	90.088		

Now the standard error of the observed difference of 0.86 oz between yarns is equal to $\sqrt{\{(\sigma_0^2 + 4\sigma_1^2)/12\}}$, and can be estimated directly from the mean square between bobbins as $\sqrt{(4.08/12)} = 0.58$ oz with 10 degrees of freedom. Thus no evidence of any systematic difference in strength between yarns A and B can be established. Confidence limits at any desired level are readily found.

Estimates of the components of variation σ_0^2, within bobbins, and σ_1^2, between bobbins, are required if, for example, recommendations are to be made for further experimentation. Suppose m bobbins are to be selected for each yarn, with l lengths per bobbin tested and that it takes k, a known value, times as long to sample a bobbin as it does to perform a single test. The total time taken is proportional to

$$T = m(k + l)$$

and the variance of the observed difference in mean strength will be proportional to

$$V = \frac{\sigma_0^2}{ml} + \frac{\sigma_1^2}{m}.$$

The recommended optimum choice of l is obtained by minimizing V, subject to given T (or minimizing T subject to given V), i.e.

$$l_{opt} = k^{\frac{1}{2}} \sigma_0 / \sigma_1.$$

Equating observed mean squares in Table Q.4 to their expected values gives estimates for the components of variance, $\tilde{\sigma}_0^2 = 1.12$ and $\tilde{\sigma}_1^2 = (4.08 - 1.12)/4 = 0.74$. Thus if, say, $k = 4$ the estimated value of l_{opt} is 2.46; we would take $l = 3$.

Further points and exercises

(i) Show that for the normal-theory estimates of variance, s^2 say, based on n observations, $\mathrm{var}(\log s^2) \simeq 2/(n-1)$. Use this result to check the homogeneity within yarns of the mean squares given in Table Q.3.

(ii) Show that the expected value of the mean square between bobbins is $\sigma_0^2 + 4\sigma_1^2$.

(iii) Determine a confidence interval for σ_1/σ_2 and hence a confidence interval for l_{opt}.

Example R

Biochemical experiment on the blood of mice

Description of data. * In an experiment on the effect of treatments A and B on the amount of substance S in mice's blood, it was not practicable to use more than 4 mice on any one day. The treatments formed a 2×2 system:

$$A_0: \text{A absent}, \qquad B_0: \text{B absent},$$
$$A_1: \text{A present}, \qquad B_1: \text{B present}.$$

The mice used on one day were all of the same sex. The data are given in Table R.1 (Cox, 1958, §7.4).

Table R.1. Amount of substance S

Day 1	Male	A_0B_1	4.8	A_1B_1	6.8	A_0B_0	4.4	A_1B_0	2.8
2	Male	A_0B_0	5.3	A_1B_0	3.3	A_0B_1	1.9	A_1B_1	8.7
3	Female	A_1B_1	7.2	A_0B_1	4.3	A_0B_0	5.3	A_1B_0	7.0
4	Male	A_0B_0	1.8	A_1B_1	4.8	A_1B_0	2.6	A_0B_1	3.1
5	Female	A_1B_1	5.1	A_0B_0	3.7	A_1B_0	5.9	A_0B_1	6.2
6	Female	A_1B_0	5.4	A_0B_1	5.7	A_1B_1	6.7	A_0B_0	6.5
7	Male	A_0B_1	6.2	A_1B_1	9.3	A_0B_0	5.4	A_1B_0	6.9
8	Female	A_0B_0	5.2	A_1B_1	7.9	A_1B_0	6.8	A_0B_1	7.9

General considerations. The experimental layout in Table R.1 is that of a split-plot design. The advantages of such a design are that the effects of subplot treatments, A and B, and their interaction with the whole-plot treatment, sex, can be estimated more precisely than can the main effect of sex, which is assumed to be of no major direct interest in the experiment. The design necessitates two different estimates of error.

The key first step in the interpretation of factorial systems such as this is the calculation and inspection of mean responses corresponding to the various combinations. Further analysis is concerned partly with assigning standard errors to the resulting contrasts and partly with finding simple summaries of the many comparisons possible.

Comparisons of treatments within males, or within females, are independent of systematic effects between days. By contrast, comparison of a particular

* Fictitious data based on a real investigation.

135

treatment across the sexes compares observations on days 1, 2, 4 and 7 with those on days 3, 5, 6 and 8, and the precision will be determined by the variation between days. The precision of comparison across the sexes has been sacrificed in the design of the experiment in favour of greater precision of comparison between treatments A and B.

In interpreting the contrasts, a distinction should be drawn between A and B, which are treatments imposed by the experimenter, and sex, which represents a natural classification of the experimental units. If there is evidence of interaction between sex and the other factors, it will be natural to look at the treatment effects separately for males and for females. It would, however, not normally be helpful to divide the data into two portions by the level of factor A and to interpret the B–sex combinations separately for the two portions.

A further general question concerns the testing of interactions for significance. Interpretation is, of course, much simpler in the absence of interactions; only marginal means need to be thought about. For the reason outlined above, distinctive attention needs to be given to the interaction with sex, this having three single degree of freedom components, A × sex, B × sex, A × B × sex. Now it turns out in the present case, that while all three components have mean squares exceeding the error mean square, only one, the last, would on its own be regarded as evidence of interaction, being significant at about the 2 per cent level. Because there has been in a sense a selection of this contrast as the most significant out of (at least) three, there is a danger of over-interpretation; the combined treatment × sex interaction with (3, 18) degrees of freedom is significant only at about the 7 per cent level.

In presenting conclusions, the reasonably cautious procedure in such cases is to give two summary conclusions. Existence of sex × treatment interactions is not firmly established and in the absence of such interaction, average effects over males and females are relevant. Nevertheless, the interaction with sex is suggestive, and if real likely to be important, so that summary conclusions separately for males and females are needed too. Of course, it might happen that external knowledge not now available to us would indicate the interactions as plausible or implausible on general grounds and this would point to the preferred interpretation. Nevertheless, the procedure we recommend would make it clear that the data are indecisive on the question of interaction.

It is wise to have some check that the conclusions are not unduly influenced by one or two extreme observations and that the conclusions would not be better expressed on a transformed scale. Because there are fairly large yet uninteresting differences between days, the simplest check is probably careful inspection of a table of day means and partial residuals eliminating day effects, i.e. differences from the day means.

The analysis. Table R.2 shows the data rearranged according to subplot treatments (A, B) and sex. For males the mean response increases only if both A and

Table R.2. Data rearranged according to treatments

Sex	Day	A_0B_0	A_1B_0	A_0B_1	A_1B_1
Male	1	4.4	2.8	4.8	6.8
	2	5.3	3.3	1.9	8.7
	4	1.8	2.6	3.1	4.8
	7	5.4	6.9	6.2	9.3
	Mean	4.22	3.90	4.00	7.40
Female	3	5.3	7.0	4.3	7.2
	5	3.7	5.9	6.2	5.1
	6	6.5	5.4	5.7	6.7
	8	5.2	6.8	7.9	7.9
	Mean	5.18	6.28	6.02	6.72
Overall mean		4.70	5.09	5.01	7.06

Estimated s.e. for difference of two overall treatment means 0.82 (18 d.f.).

B are present, whereas for females it increases in the presence of either A or B.

An analysis of variance of the data is given in Table R.3. The high variation between days within sex (m.s. = 6.06 with 6 d.f.) compared with the variation within days (m.s. = 1.33 with 18 d.f.) emphasizes the advantage of the experimental design. The combined treatments × sex interaction is only suggestive ($F = 3.76/1.33 = 2.83$ with (3, 18) d.f.) but the significance of the A × B × sex interaction ($F = 8.51/1.33 = 6.40$ with (1, 18) d.f.) is confirmed.

Table R.3. Analysis of variance

	d.f.	s.s.	m.s.
Between sexes	1	10.93	
Between days within sex	6	36.33	6.06
Between days	7	47.26	
A	1	11.88	
B	1	10.47	
A × B	1	5.53	
Treatments	3	27.88	9.29
A × Sex	1	0.81	
B × Sex	1	1.95	
A × B × Sex	1	8.51	
Treatments × Sex	3	11.27	3.76
Residual	18	24.02	1.33
Total	31	110.43	

If the treatments × sex interaction is assumed in reality to be nonexistent, and we average over the sexes, the A × B interaction is significant at a level slightly above 5 per cent ($F = 5.53/1.33 = 4.16$ with (1, 18) d.f.). Our conclusions are then summarized by the mean responses given in the bottom row of Table R.2 and the estimated standard error for the difference between any two of these means is given by $\tilde{\sigma}_0/2 = 0.578$ with 18 degrees of freedom, where $\tilde{\sigma}_0^2 = 1.33$ is the estimated variance of an observation within a particular day.

If the treatments × sex interaction is assumed to be real, we consider contrasts within the sexes. For males the A × B interaction is estimated by $(4.22 - 3.90 - 4.00 + 7.40)/2 = 1.86$, with estimated standard error 0.578 with 18 degrees of freedom, and is significant at 0.5 per cent level. Conclusions are summarized by the mean responses for males given in Table R.2. In particular, if both A and B are present, the average amount of substance S for males is increased by $7.40 - 4.22 = 3.18$ (est. s.e. 0.817 with 18 d.f.). For females the data are consistent with the absence of any A × B interaction. The estimated increase for females due to the presence of A is 0.90 and of B is 0.64 (each with est. s.e. 0.817 with 18 d.f.). Thus neither A nor B separately produces a significant increase. Confidence limits follow from the standard errors.

Further points and exercises

(i) Estimate a standard error for the comparison of treatment combination A_1B_1 across the sexes.

Related references. Snedecor and Cochran (1967, §12.12) and Armitage (1971, §8.5) discuss and give examples on the analysis of split-plot designs. Cox (1958, §17.4) discusses the design of the example.

Example S

Voltage
regulator performance

Description of data. Voltage regulators fitted to private motor cars were required to operate within the range of 15.8 to 16.4 volts, and the following investigation (Desmond, 1954) was conducted to estimate the pattern of variability encountered in production. Normal procedure was for a regulator from the production line to be passed to one of a number of setting stations, where the regulator was adjusted on a test rig. These regulators then passed to one of four testing stations, where the regulator was tested, and if found to be unsatisfactory, it was passed down the production line to be reset. For the data of Table S.1, a random sample of four setting stations took part, and a number of regulators from each setting station were passed through each testing station. One special aspect of interest concerned the percentage of regulators that would be unsatisfactory were the mean kept constant at 16.1.

General considerations. These data have a fairly highly balanced structure with a number of explanatory variables (factors) and a quantitative response varying over quite a narrow range. The powerful techniques of analysis of variance are thus available, but careful inspection for heterogeneity, outliers, etc., is wise.

The primary objective is to isolate that part of the observed variation that is 'real' variation between regulators, as contrasted with testing or measurement error. For this the technique of analysis of variance, in the literal sense of breaking variance into components, is natural, followed by synthesis of variance, i.e. the reconstruction of the variance associated with the part of the variation that is of concern. The analysis of variance does not require normality; on the other hand, the specific question posed does require the approximate normality of the relevant portions of the variability. This can be checked to a limited extent, and with very extensive data more elaborate methods might be used, involving, for example, the estimation of third and fourth cumulants.

The role of the analysis-of-variance table is:

(i) as a concise summary of the data;

(ii) as an indicator of which sources of variation, especially interaction, may be assumed absent;

139

Table S.1. Regulator voltages

Setting station	Regulator number	Testing station 1	2	3	4	Setting station	Regulator number	Testing station 1	2	3	4
A	1	16.5	16.5	16.6	16.6	F	1	16.1	16.0	16.0	16.2
	2	15.8	16.7	16.2	16.3		2	16.5	16.1	16.5	16.7
	3	16.2	16.5	15.8	16.1		3	16.2	17.0	16.4	16.7
	4	16.3	16.5	16.3	16.6		4	15.8	16.1	16.2	16.2
	5	16.2	16.1	16.3	16.5		5	16.2	16.1	16.4	16.2
	6	16.9	17.0	17.0	17.0		6	16.0	16.2	16.2	16.1
	7	16.0	16.2	16.0	16.0		11	16.0	16.0	16.1	16.0
	11	16.0	16.0	16.1	16.0	G	1	15.5	15.5	15.3	15.6
							2	16.0	15.6	15.7	16.2
B	1	16.0	16.1	16.0	16.1		3	16.0	16.4	16.2	16.2
	2	15.4	16.4	16.8	16.7		4	15.8	16.5	16.2	16.2
	3	16.1	16.4	16.3	16.3		5	15.9	16.1	15.9	16.0
	4	15.9	16.1	16.0	16.0		6	15.9	16.1	15.8	15.7
							7	16.0	16.4	16.0	16.0
C	1	16.0	16.0	15.9	16.3		12	16.1	16.2	16.2	16.1
	2	15.8	16.0	16.3	16.0	H	1	15.5	15.6	15.4	15.8
	3	15.7	16.2	15.3	15.8		2	15.8	16.2	16.0	16.2
	4	16.2	16.4	16.4	16.6		3	16.2	15.4	16.1	16.3
	5	16.0	16.1	16.0	15.9		4	16.1	16.2	16.0	16.1
	6	16.1	16.1	16.1	16.1		5	16.1	16.2	16.3	16.2
	10	16.1	16.0	16.1	16.0		10	16.1	16.1	16.0	16.1
						J	1	16.2	16.1	15.8	16.0
D	1	16.1	16.0	16.0	16.1		2	16.2	15.3	17.8	16.3
	2	16.0	15.9	16.2	16.0		3	16.4	16.7	16.5	16.5
	3	15.7	15.8	15.7	15.7		4	16.2	16.5	16.1	16.1
	4	15.6	16.4	16.1	16.2		5	16.1	16.4	16.1	16.3
	5	16.0	16.2	16.1	16.1		10	16.4	16.3	16.4	16.4
	6	15.7	15.7	15.7	15.7	K	1	15.9	16.0	15.8	16.1
	11	16.1	16.1	16.1	16.0		2	15.8	15.7	16.7	16.3
							3	16.2	16.2	16.2	16.3
E	1	15.9	16.0	16.0	16.5		4	16.2	16.3	15.9	16.3
	2	16.1	16.3	16.0	16.0		5	16.0	16.0	16.0	16.0
	3	16.0	16.2	16.0	16.1		6	16.0	16.4	16.2	16.2
	4	16.3	16.5	16.4	16.4		11	16.0	16.1	16.0	16.1

(iii) as a basis for significance tests making (ii) more precise;

(iv) as a basis for the estimation of components of variance.

The analysis. Inspection across the rows of Table S.1 shows for most regulators close consistency in the four readings. Exceptions are regulators B_2 (15.4, 16.4, 16.8, 16.7) and J_2 (16.2, 15.3, 17.8, 16.3). Examination of residuals from the two-way tables, regulators × testing stations, within each setting station, confirms the high variation across J_2; regulators, B_2, D_4, F_2, H_3, and K_2 also have less consistent observations than the majority, although not to the extent of J_2. To check on the influence of J_2 on the final conclusions, two parallel analyses have been done, one with and one without J_2. Of course, had we been in a position to look into any special circumstances connected with this regulator, we would have done so.

Table S.2. Residual mean squares
within setting stations

Setting station	d.f.	m.s.
A	21	0.0319
B	9	0.0869
C	18	0.0363
D	18	0.0209
E	9	0.0245
F	18	0.0357
G	21	0.0306
H	15	0.0423
J (full data)	15	0.2227
(J_2 omitted)	12	0.0160
K	18	0.0445

It aids in examining homogeneity to begin with a separate two-way analysis of variance for each of the 10 setting stations. Table S.2 gives the residual mean squares. The high value for setting station J is attributable to regulator J_2; the omission of J_2 reduces the mean square from 0.2227 to 0.0160, in reasonable accord with the other setting stations.

Pooling the ten analyses leads to the analysis of variance in Table S.3, shown in two versions with and without J_2. An initial conclusion is that the interaction term, setting stations × testing stations, is accountable by random error. We can thus regard each observation as deviating from a notional 'true' value for that regulator, holding if setting-station effects were eliminated, by the sum of a testing-station contribution, a setting-station contribution and a random term. The former is irrelevant for the present purpose. If we denote by σ_s^2 the component of variance for setting station effects, σ_r^2 the variance between regulators of the 'true' values and σ_0^2 the error variance, the expected mean squares are as shown in Table S.3.

Table S.3. Analysis of variance

	Full data		J_2 omitted		E(m.s.)
	d.f.	m.s.	d.f.	m.s.	
Setting stations (SS)	9	0.4910	9	0.4625	$\sigma_0^2 + 4\sigma_r^2 + \dfrac{4}{9}\left(\sum m_i - \dfrac{\sum m_i^2}{\sum m_i}\right)\sigma_s^2$
Testing stations (TS)	3	0.2615	3	0.2985	
SS × TS	27	0.0335	27	0.0267	
Regulators within SS	54	0.1756	53	0.1779	$\sigma_0^2 + 4\sigma_r^2$
Residual	162	0.0541	159	0.0353	σ_0^2

m_i is the number of regulators from setting station i

Now as produced on a particular setting-station, regulator 'true' voltage has standard deviation $\sqrt{(\sigma_s^2 + \sigma_r^2)}$. This can be estimated via Table S.3 and the relevant estimates are given in Table S.4. Exclusion of J_2 reduces $\tilde{\sigma}_0^2$ but has little effect upon $\tilde{\sigma}_s^2$ or $\tilde{\sigma}_r^2$. Thus if J_2 is included, we estimate the standard deviation as $\sqrt{(\tilde{\sigma}_s^2 + \tilde{\sigma}_r^2)} = 0.207$; if J_2 is excluded this becomes 0.217.

Graphical plots confirm the assumption of normality of the between-regulator and between-setting station variation. Thus with the mean fixed at 16.1, the estimated percentage falling outside 15.8–16.4 volts is, assuming a normal distribution, 14.7 per cent with J_2 included, and 16.6 per cent if J_2 is excluded. While these estimates are based on the normal distribution, the use of moderately non-normal distributions would not change the values appreciably (Pearson and Hartley, 1972, Table 32).

Assuming normality of the between-regulator and between-setting station variation, approximate confidence intervals for the percentage outside the tolerance limits are calculated by taking $(\tilde{\sigma}_s^2 + \tilde{\sigma}_r^2)$ to be approximately proportional to χ^2, with 'effective' degrees of freedom determined by adjusting the

Table S.4. Estimated components of variance

	Full data	J_2 omitted
$\tilde{\sigma}_0^2$	0.0541	0.0353
$\tilde{\sigma}_r^2$	0.0304	0.0357
$\tilde{\sigma}_s^2$	0.0124	0.0114

mean and variance. If J_2 is included this gives 'effective' degrees of freedom approximately 27, and approximate 95 per cent confidence limits for $\sqrt{(\sigma_s^2 + \sigma_r^2)}$ of 0.163 to 0.281 with corresponding limits for the percentage of regulators outside the tolerance limits of 6.6 to 28.6 per cent. If J_2 is excluded these limits become 8.7 to 29.0 per cent. The effect of omitting J_2 is unimportant in the light of the precision of the analysis.

Further points and exercises

(i) Write down an appropriate model for the observations and check the expected mean squares quoted in Table S.3.

(ii) Using the suggested χ^2 approximation, check that $(\tilde{\sigma}_s^2 + \tilde{\sigma}_r^2)$ has 'effective' degrees of freedom approximately 27 and thence confirm the stated confidence limits for $\sqrt{(\sigma_s^2 + \sigma_r^2)}$.

Related reference. Snedecor and Cochran (1967, §12.11) describe the use of 'effective' degrees of freedom in components-of-variance problems.

Example T   **Intervals between**
the failure of air-conditioning
equipment in aircraft

Description of data. The data in Table T.1, reported by Proschan (1963), are the intervals in service-hours between failures of the air-conditioning equipment in 10 Boeing 720 jet aircraft. It is required to describe concisely the variation within and between aircraft, with emphasis on the forms of the frequency distributions involved.

General considerations. The data are a special form of time series. A wide variety of aspects may be explored and it is necessary therefore to be guided to some extent by the practical object in mind. The following are among the matters that could be considered:

(i) Do failures depend on the time of year, or more generally on external explanatory variables? The information to tackle this is not available in the present instance.

(ii) Do failure intervals on a particular aircraft vary randomly, or is there a trend or serial correlation in their values?

(iii) Assuming stationarity and independence, is the frequency distribution of intervals for any one aircraft essentially exponential, corresponding to a Poisson process or completely random set of point occurrences? If not, is there a simple qualitative or quantitative description of the departure from exponential form, for example via the fitting of a gamma or Weibull distribution?

(iv) Are the distributions for the different aircraft the same, and if not how can the differences between aircraft be described concisely?

All these questions can be tackled by methods ranging from the informal and graphical to the fitting by maximum likelihood of plausible parametric models. We concentrate on (iii) and (iv), regarding absence of trends and serial correlation as secondary issues adequately checked by simple graphical techniques.

Some methodological details. There are many ways of checking agreement with exponential form. The most useful graphical technique is probably to plot the ordered times

143

Table T.1. Intervals between failures (operating hours)

Aircraft number									
1	2	3	4	5	6	7	8	9	10
413	90	74	55	23	97	50	359	487	102
14	10	57	320	261	51	44	9	18	209
58	60	48	56	87	11	102	12	100	14
37	186	29	104	7	4	72	270	7	57
100	61	502	220	120	141	22	603	98	54
65	49	12	239	14	18	39	3	5	32
9	14	70	47	62	142	3	104	85	67
169	24	21	246	47	68	15	2	91	59
447	56	29	176	225	77	197	438	43	134
184	20	386	182	71	80	188		230	152
36	79	59	33	246	1	79		3	27
201	84	27	15	21	16	88		130	14
118	44	153	104	42	106	46			230
34	59	26	35	20	206	5			66
31	29	326		5	82	5			61
18	118			12	54	36			34
18	25			120	31	22			
67	156			11	216	139			
57	310			3	46	210			
62	76			14	111	97			
7	26			71	39	30			
22	44			11	63	23			
34	23			14	18	13			
	62			11	191	14			
	130			16	18				
	208			90	163				
	70			1	24				
	101			16					
	208			52					
				95					

$$Y_{(1n)} \leq Y_{(2n)} \leq \ldots \leq Y_{(nn)}$$

against the expected order statistics in sampling the exponential distribution of unit mean, namely

$$e_{(1n)} < e_{(2n)} < \ldots < e_{(nn)},$$

where $e_{(1n)} = 1/n$, $e_{(2n)} = 1/n + 1/(n-1), \ldots, e_{(nn)} = 1/n + 1/(n-1) + \ldots + 1$. Departure from an exponential distribution is shown by systematic non-linearity.

If the intervals on each aircraft are assumed to be exponentially distributed, a comparison between aircraft is a comparison of sample means. One test of equality of means in a version of Bartlett's test for homogeneity of variance, i.e. uses twice a difference of maximized log likelihoods. The test statistic is

$$-2N \log(Y/N) + 2 \, \Sigma \, n_i \log(Y_i/n_i), \tag{T.1}$$

where n_i is the number of observations for aircraft i ($i = 1, \ldots, k$), $Y_i = \Sigma_j Y_{ij}$, $Y = \Sigma_i Y_i$, $N = \Sigma_i n_i$. The distribution under the null hypothesis is approximately χ^2 with $k-1$ degrees of freedom.

A more formal procedure is to fit a gamma distribution,

$$\frac{(\beta/\mu)(\beta y/\mu)^{\beta-1} e^{-\beta y/\mu}}{\Gamma(\beta)} \tag{T.2}$$

to the results for each aircraft. A rather systematic procedure is to fit by maximum likelihood the following models:

(i) separate gamma distributions to all aircraft, with 20 parameters;
(ii) separate gamma distributions with a common β, with 11 parameters;
(ii) common gamma distribution to all aircraft, with 2 parameters;
(iv) separate exponential distributions to all aircraft ($\beta = 1$, separate μ), with 10 parameters;
(v) common exponential distribution to all aircraft ($\beta = 1$), with 1 parameter.

All the fittings are essentially straightforward and comparison of the maximized log likelihoods allows various approximate significance tests to be made.

The analysis. Fitting separate gamma distributions to each aircraft leads to the estimated values of μ and β given in Table T.2. If we test the hypothesis of a common value of β, direct comparison of the maximum log likelihoods achieved gives a value of χ^2 of 31.50 with 9 degrees of freedom, i.e. differences are highly significant.

Unfortunately, the interpretation is not straightforward. One aircraft, namely no. 8, has a very low value of $\hat{\beta}$, i.e. very high variability. This is not accounted for by one or two possibly anomalous observations. Next, if the data from this aircraft are omitted and the remaining aircraft analysed there is still appreciable dispersion in the individual $\hat{\beta}$'s, the value of χ^2 being now 19.55 with 8 degrees of freedom.

Table T.2. Maximum-likelihood estimates of mean μ and index β

	Aircraft									
	1	2	3	4	5	6	7	8	9	10
$\hat{\mu}$	95.7	83.5	121.3	130.9	59.6	76.8	64.1	200.0	108.1	82.0
$\hat{\beta}$	0.97	1.67	0.83	1.61	0.81	1.13	1.06	0.46	0.71	1.75

Now the use of the chi-squared distribution for assessing differences of maximized log likelihoods is based on a mathematical approximation and usually tends to overestimate the evidence against the null hypothesis. In the present case a more refined calculation changes 19.55 to 16.84; the tabulated 5 per cent and 1 per cent points are respectively 15.51 and 20.09. Thus, even after modification, there is too much variation among the β's to be reasonably accounted for by random fluctuations. If, nevertheless, a common value of β is assumed, its maximum-likelihood estimate is 1.07, very close to the value 1 for an exponential distribution.

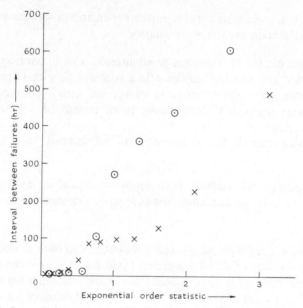

Fig. T.1. Intervals between successive failures versus exponential order statistics.
⊙   Aircraft 8
✕   Aircraft 9

Figure T.1 shows the data, for the two aircraft (numbers 8 and 9) with the smallest β's, plotted against exponential order statistics. Both aircraft show a high proportion of small values but otherwise suggest irregular variation, rather than any systematic departure from an exponential distribution.

This illustrates a fairly common dilemma in applications. There is a very simple representation of the data, here based on exponential distributions, that would make interpretation vivid. In some average sense the fit is reasonable, but there are certainly unexplained deviations. First one should try to explain the discrepancies, i.e. the variation in $\hat{\beta}$. There is no suggestion that $\hat{\beta}$ is related to the mean and we do not have other information about the individual aircraft on which to base an explanation. It is possible that the variation is in some sense unreal, a fairly extreme random fluctuation, conceivably amplified by any positive correlation between intervals within aircraft. Whether, nevertheless, the exponential distribution should be the basis of future applications depends on the purpose. If occurrences of very short or very long intervals are of concern, it would probably be unwise to use exponential distributions: if main interest lies in means, their standard errors and in rough extrapolation into the tails of the distribution, use of exponential distributions is likely to be sensible. Of course, any report on the conclusions must include a statement about the unexplained discrepancies.

Subject to these reservations, we can assume exponential distributions and test for a common failure rate using the test statistic (T.1). This gives $\chi^2 = 19.7$ with 9 degrees of freedom, which is significant at about the $2\frac{1}{2}$ per cent level. If aircraft 8 is excluded, this becomes $\chi^2 = 11.7$ with 8 degrees of freedom and is not significant, confirming that aircraft 8, with a mean of 200.0 hours, has a significantly low failure rate.

Related reference. Cox and Lewis (1966) discuss analysis of these data, except that here we have excluded three aircraft with very few observations.

Example U

Survival times of leukemia patients

Description of data. The data in Table U.1 from Feigl and Zelen (1965) are time to death, Y, in weeks from diagnosis and $\log_{10}$ (initial white blood cell count), x, for 17 patients suffering from leukemia. The relation between Y and x is the main aspect of interest.

Table U.1. Survival time Y in weeks and $\log_{10}$ (initial white blood cell count) for 17 leukemia patients

x	Y	x	Y	x	Y
3.36	65	4.00	121	4.54	22
2.88	156	4.23	4	5.00	1
3.63	100	3.73	39	5.00	1
3.41	134	3.85	143	4.72	5
3.78	16	3.97	56	5.00	65
4.02	108	4.51	26		

General considerations. A plot of survival time, Y, versus log (white blood cell count), x, shows substantial random variation together with a tendency for Y to decrease with increasing x (see Fig. U.1). Elaborate model fitting would be unnecessary for the analysis of these data in isolation. In particular, many different parametric representations of the systematic variations are consistent with the data.

We use the data to illustrate two main points. One is the use of general considerations to choose between alternative parametric regression relations. The other is the examination of the form of the random variation about such a relation.

Three main aspects are involved in setting up a parametric description:

(i) a specification of the form of systematic variation, for example by giving the relation between the expected value $E(Y_i)$ and x_i for the ith individual;

(ii) a specification of the general form of the random variation;

(iii) a specification of the way in which systematic and random variation 'combine', e.g. by multiplication or by addition.

148

It is preferable for (i) to choose where possible a relation that gives positive values for $E(Y)$ for all possible parameter values and values of x. From this point of view, the relation

$$E(Y_i) = \beta_0 \exp\{\beta_1(x_i - \bar{x})\},$$

where $\bar{x} = \Sigma\, x_i/n$, is preferable to, for example, a linear relation

$$E(Y_i) = \gamma_0 + \gamma_1(x_i - \bar{x}).$$

If the random contribution to the ith observation is ε_i, simplicity of interpretation and fitting, and inspection of the data, suggest a multiplicative combination; that is, we consider an interpretation in which the proportional variation around the mean has the same form for all x. It is then of interest to compare that distribution with the exponential distribution, partly because that is about the simplest very dispersed distribution for a positive quantity, partly because of the interpretation of the exponential distribution in terms of the properties of a completely random process, the Poisson process, and partly because use of the exponential distribution much simplifies more detailed analysis and consistency with exponential form is of general interest.

These considerations lead to the model

$$Y_i = \beta_0 \exp\{\beta_1(x_i - \bar{x})\}\varepsilon_i, \tag{U.1}$$

where ε_i is a random term of unit mean and, conceivably, exponentially distributed.

Some methodological details. To fit the model (U.1), assuming an exponential distribution for the random components, iterative solution of the maximum-likelihood equations is the most effective procedure, leading to estimates $\hat{\beta}_0$ and $\hat{\beta}_1$. We can then define residuals which are in effect 'estimates' of the corresponding errors. For the ith observation we put

$$R_i = Y_i[\hat{\beta}_0 \exp\{\hat{\beta}_1(x_i - \bar{x})\}]^{-1}, \tag{U.2}$$

the ratio of the observed to fitted survival time. The R_i can be examined for distributional form, by plotting against expected exponential order statistics; see Example T. Departure from a multiplicative relation would be shown by a relation between R_i and x_i.

The analysis. For the model (U.1), in which ε_i has a unit exponential distribution, the equations satisfied by the maximum-likelihood estimates $\hat{\beta}_0$ and $\hat{\beta}_1$ can be written as

$$n\hat{\beta}_0 = \Sigma\, Y_i \exp\{-\hat{\beta}_1(x_i - \bar{x})\}, \tag{U.3}$$

$$0 = \Sigma\, Y_i(x_i - \bar{x}) \exp\{-\hat{\beta}_1(x_i - \bar{x})\}, \tag{U.4}$$

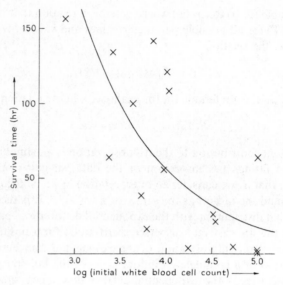

Fig. U.1. Survival time versus $\log_{10}$ (initial white blood cell count).
———————— fitted exponential model (U.1)

with $n = 17$. Iterative solution of Equation (U.4) leads to $\hat{\beta}_1 = -1.109$ and substitution into Equation (U.3) gives $\hat{\beta}_0 = 51.109$. The fitted model is shown superimposed in Fig. U.1.

Asymptotic standard errors are estimated to be as follows:

$$\text{est. s.e. } (\hat{\beta}_0) \simeq \hat{\beta}_0/\sqrt{n} = 12.396,$$
$$\text{est. s.e. } (\hat{\beta}_1) \simeq \{\Sigma (x_i - \bar{x})^2\}^{-\frac{1}{2}} = 0.400,$$

and $\text{cov}(\hat{\beta}_0, \hat{\beta}_1) = 0$.

The adequacy of the fitted model is confirmed if the residuals R_i as defined by Equation (U.2) are plotted against exponential order statistics as in Example T.

Further points and exercises

(i) Consider the determination of a confidence interval for the expected value of Y for a given value x.

(ii) Consider the determination of a prediction interval for the survival time of a new individual with a given value of x.

(iii) Discuss some methods of fitting alternative to maximum likelihood.

Example V

A retrospective study with binary data

Description of data. In a retrospective study of the possible effect of blood group on the incidence of peptic ulcers, Woolf (1955) obtained data from three cities. Table V.1 gives for each city data for blood groups O and A only. In each city, blood group is recorded for peptic-ulcer subjects and for a control series of individuals not having peptic ulcer.

Table V.1. Blood groups for peptic ulcer and control subjects

	Peptic ulcer		Control	
	Group O	Group A	Group O	Group A
London	911	579	4578	4219
Manchester	361	246	4532	3775
Newcastle	396	219	6598	5261

General considerations. This is an example of a type of retrospective investigation widely used, in particular in epidemiology under the name case-control study. We want really to study how the probability of peptic ulcer depends on blood group, i.e. to use occurrence of peptic ulcer as a response and blood group as an explanatory variable. For fairly obvious reasons, it is convenient to collect data in an inverse fashion, taking a set of peptic-ulcer patients and a set chosen from non-peptic ulcer individuals and then observing blood group for each individual. The possibility of using data of this type to answer the question of interest depends on the identity

$$\Delta = \log\left\{\frac{\text{pr (ulcer|A)}}{\text{pr (ulcer|O)}} \times \frac{\text{pr (no ulcer|O)}}{\text{pr (no ulcer|A)}}\right\} \qquad (V.1)$$

$$= \log\left\{\frac{\text{pr (A|ulcer)}}{\text{pr (O|ulcer)}} \times \frac{\text{pr (O|no ulcer)}}{\text{pr (A|no ulcer)}}\right\}. \qquad (V.2)$$

The second of these can be estimated from each city by replacing probabilities by proportions. We use large-sample theory, thus obtaining an estimate of Δ, with a standard error. The three estimates can be tested for consistency

151

with a common Δ and, if they are reasonably consistent, a weighted mean and confidence limits are calculated.

Finally, note that if in the population the probability of having an ulcer is fairly small, then (V.1) is effectively

$$\log\left\{\frac{\text{pr (ulcer|A)}}{\text{pr (ulcer|O)}}\right\}, \tag{V.3}$$

so that $\exp(\Delta)$ gives the ratio of occurrence probabilities in groups A and O.

Some methodological details. Suppose that in city j, samples of sizes n_{1j} and n_{2j} from control and ulcer groups give

	A	O
A	R_{1j}	R_{2j}
O	$n_{1j}-R_{1j}$	$n_{2j}-R_{2j}$
Total	n_{1j}	n_{2j}

Then an estimate of Δ_j, the parameter (V.1) for city j, is

$$\tilde{\Delta}_j = \log\left(\frac{R_{2j}}{n_{2j}-R_{2j}}\right) - \log\left(\frac{R_{1j}}{n_{1j}-R_{1j}}\right) \tag{V.4}$$

and this has large-sample variance

$$v_j = \frac{1}{R_{2j}} + \frac{1}{n_{2j}-R_{2j}} + \frac{1}{R_{1j}} + \frac{1}{n_{1j}-R_{1j}}, \tag{V.5}$$

assuming the R's to be independently binomially distributed.

Thus a χ^2 statistic for conformity with constant Δ_j is

$$\begin{aligned}
\chi^2 &= \Sigma\,(\tilde{\Delta}_j - \tilde{\Delta}.)^2/v_j \\
&= \Sigma\,\tilde{\Delta}_j^2/v_j - (\Sigma\,\tilde{\Delta}_j/v_j)^2(\Sigma\,1/v_j)^{-1},
\end{aligned} \tag{V.6}$$

where

$$\tilde{\Delta}. = \frac{\Sigma\,\tilde{\Delta}_j/v_j}{\Sigma\,1/v_j} \tag{V.7}$$

is the weighted mean with large-sample variance

$$(\Sigma 1/v_j)^{-1}. \tag{V.8}$$

If in fact all Δ_j are equal, the large-sample distribution of (V.6) is χ^2 with degrees of freedom the number of cities minus one.

The analysis. Table V.2 shows the estimates $\tilde{\Delta}_j$, their standard errors $\sqrt{v_j}$ and the estimated ratios $\exp(\tilde{\Delta}_j)$. For example, for the London data, $j = 1$,

Example V   153

$$\tilde{\Delta}_1 = \log \frac{579}{911} - \log \frac{4219}{4578} = -0.3716,$$

$$\tilde{v}_1 = 1/579 + 1/911 + 1/4219 + 1/4578,$$

etc. The χ^2 statistic (V.6) is 2.98, with two degrees of freedom; this has $P \simeq 0.2$.

Table V.2. Estimated logistic differences $\tilde{\Delta}_j$ and standard errors $\sqrt{v_j}$

	j	$\tilde{\Delta}_j$	$\sqrt{v_j}$	$\exp(\tilde{\Delta}_j)$
London	1	-0.3716	0.05727	0.700
Manchester	2	-0.2008	0.08556	0.818
Newcastle	3	-0.3659	0.08622	0.694
Pooled		-0.3298	0.04167	0.719

Thus all three cities show a lower peptic ulcer rate for group A than for group O and the variation in the ratios between cities, while a little greater than would be expected on the average by chance, is well within the limits to be expected in random sampling. The pooled estimate, the weighted mean $\tilde{\Delta}_.$, is -0.330, with standard error 0.042, the corresponding ratio of probabilities being estimated as 0.719; a 95 per cent confidence interval for the ratio is (0.663, 0.780).

The main assumption limiting the conclusions is that the control series can be treated as effectively randomly chosen from that part of the target population without peptic ulcer. Obviously, for example, if the control individuals all suffered from some disease itself associated with blood group, then a bias would be introduced. The calculations of precision assume independent individuals with constant chance of 'success', leading to binomial variation. Effective matching between ulcer and control groups might increase precision and any form of 'cluster' sampling could decrease precision, i.e. increase variance. It seems most unlikely in the present instance that the qualitative conclusions could be affected by such changes in precision.

Further points and exercises

(i) Prove the identity of (V.1) and (V.2).

(ii) Prove the formula (V.5) for the variance, using the variance of the binomial distribution for the random variables R_{1j} and R_{2j}.

(iii) How would the analysis have proceeded had the $\tilde{\Delta}_j$ for the three cities differed significantly?

(iv) Suppose that it is required to estimate $\text{pr(ulcer}|A) - \text{pr(ulcer}|O)$. What further information is needed to do this from the current data?

(v) Consider the advantages and disadvantages of the retrospective study illustrated here as compared with alternative schemes of investigation.

Related references. Armitage (1971, §§6.3, 16.2) describes the use and analysis of retrospective studies, especially in epidemiology, including the test of homogeneity. Snedecor and Cochran (1967, §16.11) explain the use of the empirical logit transform, and Cox (1970, Chapter 3) gives a more detailed discussion.

Example W
Housing and associated factors

Description of data. The data in Table W.1 (Madsen, 1976) relate to an investigation into satisfaction with housing conditions in Copenhagen. A total of 1681 residents from selected areas living in rented homes built between 1960 and 1968 were questioned on their satisfaction, the degree of contact with other residents and their feeling of influence on apartment management. The purpose of the investigation was to study association between these three factors and the type of housing.

Table W.1. 1681 persons classified according to satisfaction, contact, influence and type of housing

Contact		Low			High		
Satisfaction		Low	Medium	High	Low	Medium	High
Housing	*Influence*						
Tower blocks	Low	21	21	28	14	19	37
	Medium	34	22	36	17	23	40
	High	10	11	36	3	5	23
Apartments	Low	61	23	17	78	46	43
	Medium	43	35	40	48	45	86
	High	26	18	54	15	25	62
Atrium houses	Low	13	9	10	20	23	20
	Medium	8	8	12	10	22	24
	High	6	7	9	7	10	21
Terraced houses	Low	18	6	7	57	23	13
	Medium	15	13	13	31	21	13
	High	7	5	11	5	6	13

Some general considerations. In observational studies such as this the distinction between response and explanatory variables is not so clear as in controlled experiments. It is reasonable at a descriptive level to take type of housing as an explanatory variable, although even here care is needed in ultimate interpretation, in that we do not know 'why' individuals live in one type of housing rather than another, and differences in response between

155

different types of housing may well reflect some deeper explanatory variable.

For some purposes it might be reasonable to take satisfaction as the single response variable to be 'explained' in terms of the other variables: influence and contact would then be intermediate variables. We shall, however, here concentrate on an analysis in which satisfaction, influence and contact are treated symmetrically as three response variables whose variation is to be described in as simple a way as possible. That is, we treat the investigation as having three response variables and one explanatory variable.

As in multivariate problems generally, it is sensible to begin with the response variables one at a time. For this we examine for each type of housing the marginal frequencies of the various categories of response, examining each response variable separately. Then we turn to the association between the response variables. The simplest, but unlikely, possibility is that for each type of housing the three types of response are independent of one another. This is not the case here. Next it might happen that one of the response variables is independent of the other two, possibly with the pattern of association between the last two variables being similar for the different types of housing.

The fitting of log linear models is one powerful tool for exploring such matters. They are computationally more difficult to handle than the corresponding linear models for approximately normally distributed data, but this is unimportant provided a suitable computer program such as GLIM is available. A more serious difficulty with models lies in its sometimes being hard to interpret the model or models ultimately chosen as best representing the data. One important aspect is that forming, say, two-way tables of satisfaction versus contact, for a particular type of housing, involves *addition* of frequencies over the third variable influence. Yet the log linear models are specified multiplicatively, essentially because independence implies multiplication of marginal probabilities. Thus a parameter fitted to represent, say, a portion of the above two-factor interaction, allowing for the possible presence of two-factor interactions between the other variables, refers to the supposed relation between satisfaction and contact, for a fixed level of influence, and not, directly at least, to the two-way table of satisfaction–contact frequencies.

In practice, the most important use of the models is likely to be to indicate the level of complexity needed to represent the data: having settled this, it will be important to present the conclusions in as direct a form as possible and this will often be via tables of fitted frequencies implied by the model rather than by the estimated parameters themselves.

As with analysis of relatively complex data in general, some choice is needed between starting with fits to the whole data, attempting to achieve reasonable simplification of the most general model, and between separate fits to rational sections of the data, in this case the sections being the different types of housing. The choice depends partly on the analyst's experience with the material. Here we have put some emphasis on the second approach of starting with sections. It is important in doing this, however, to aim as far as possible at broadly similar representations of the different sections.

The analysis. Table W.2 shows for each type of housing the distribution of residents according to their responses on satisfaction, degree of contact and feeling of influence. There are marked differences in distribution between types of housing. Satisfaction is highest in tower blocks (50% responding 'high') and lowest in terraced houses (25% responding 'high'), contact is highest in atrium (66% 'high') and terraced houses (66% 'high'), influence is lowest in terraced houses (45% 'low').

Such marginal tables have to be interpreted carefully. That for satisfaction, for instance, describes the distribution of that variable on its own for each

Table W.2. Distribution of respondents according to satisfaction, contact and influence for each type of housing

Type of housing	Satisfaction			
	Low (%)	Medium (%)	High (%)	Total no. respondents
Tower blocks	24.8	25.2	50.0	400
Apartments	35.4	25.1	39.5	765
Atrium houses	26.8	33.1	40.2	239
Terraced houses	48.0	26.7	25.3	277

	Contact		
	Low (%)	High (%)	
Tower blocks	54.8	45.2	400
Apartments	41.4	58.6	765
Atrium houses	34.3	65.7	239
Terraced houses	34.3	65.7	277

	Influence			
	Low (%)	Medium (%)	High (%)	
Tower blocks	35.0	43.0	22.0	400
Apartments	35.0	38.8	26.1	765
Atrium houses	39.7	35.1	25.1	239
Terraced houses	44.8	38.3	17.0	277

Table W.3. Approximate χ^2 for log linear models fitted to each type of housing

Housing		Tower blocks	Apartments	Atrium houses	Terraced houses
Model	d.f.	Approximate χ^2			
(a) Main effects	12	28.8	98.0	11.6	40.4
(b) Main effects plus $C \times S$	10	22.1	90.2	9.1	36.2
$C \times I$	10	23.8	91.4	11.4	30.2
$S \times I$	8	14.3	22.4	3.9	14.2

Approximate χ^2 is twice the difference of maximized log likelihood for full model (perfect fit) and model indicated. In GLIM it is called deviance.

type of housing and as such is of direct interest, at least descriptively. Yet when the response variables are associated it could happen that for any particular level of influence the conditional distribution of satisfaction is the same for all types of housing and yet the marginal distribution may differ.

The results of fitting log linear models to the three response variables, separately for each type of housing, are summarized in Table W.3, giving an approximate χ^2 as twice the difference of log maximized likelihood between the fitted model and the full model giving a perfect fit. In GLIM this is called deviance. Tower blocks, atrium houses and terraced houses need only one interaction $S \times I$ (satisfaction $\times$ influence) to explain the data; the corresponding values of χ^2, each with 8 degrees of freedom, are 14.3, 3.9 and 14.2,

Table W.4. Residuals from model containing interaction $S \times I$, within type of housing

Contact		Low			High		
Satisfaction		L	M	H	L	M	H
Housing Tower blocks	Influence						
	L	0.4	−0.2	−1.3	−0.5	0.2	1.4
	M	1.2	−0.5	−0.9	−1.3	0.6	1.0
	H	1.1	0.8	0.7	−1.2	−0.8	−0.7
Apartments							
	L	0.4	−1.0	−1.6	−0.4	0.9	1.3
	M	0.9	0.3	−1.7	−0.7	−0.3	1.4
	H	2.2	0.0	0.9	−1.8	0.0	−0.7
Atrium houses							
	L	0.5	−0.6	−0.1	−0.4	0.4	0.1
	M	0.7	−0.7	−0.1	−0.5	0.5	0.1
	H	0.7	0.5	−0.4	−0.5	−0.3	0.3
Terraced houses							
	L	−1.5	−1.3	0.1	1.1	0.9	0.0
	M	−0.2	0.4	1.4	0.1	−0.3	−1.0
	H	1.4	0.6	1.0	−1.0	−0.5	−0.7

| Contact | Low | | | | | | High | | | | | |
| Satisfaction | L | | M | | H | | L | | M | | H | |
Housing	Influence	o	f	o	f	o	f	o	f	o	f	o	f
Tower blocks	L	21	19.2 / 19.0	21	21.9 / 19.4	28	35.6 / 38.3	14	15.8 / 15.7	19	18.1 / 16.0	37	29.4 / 31.7
	M	34	27.9 / 23.3	22	24.6 / 23.8	36	41.6 / 47.1	17	23.1 / 19.3	23	20.4 / 19.7	40	34.4 / 38.9
	H	10	7.1 / 11.9	11	8.8 / 12.2	36	32.3 / 24.1	3	5.9 / 9.9	5	7.2 / 10.1	23	26.7 / 19.9
Apartments	L	61	57.6 / 39.3	23	28.6 / 27.9	17	24.9 / 43.8	78	81.4 / 55.6	46	40.4 / 39.4	43	35.1 / 62.0
	M	43	37.7 / 43.6	35	33.2 / 30.9	40	52.2 / 48.6	48	53.3 / 61.6	45	46.9 / 43.7	86	73.8 / 68.7
	H	26	17.0 / 29.4	18	17.8 / 20.8	54	48.1 / 32.7	15	24.0 / 41.5	25	25.2 / 29.4	62	67.9 / 46.2
Atrium houses	L	13	11.3 / 8.7	9	11.0 / 10.8	10	10.3 / 13.1	20	21.7 / 16.7	23	21.0 / 20.6	20	19.7 / 25.1
	M	8	6.2 / 7.7	8	10.3 / 9.5	12	12.4 / 11.6	10	11.8 / 14.8	22	19.7 / 18.2	24	23.7 / 22.2
	H	6	4.5 / 5.5	7	5.8 / 6.8	9	10.3 / 8.3	7	8.5 / 10.6	10	11.2 / 13.0	21	19.7 / 15.8
Terraced houses	L	18	25.7 / 20.4	6	9.9 / 11.4	7	6.9 / 10.8	57	49.3 / 39.1	23	19.1 / 21.8	13	13.1 / 20.6
	M	15	15.8 / 17.5	13	11.7 / 9.7	13	8.9 / 9.2	31	30.2 / 33.4	21	22.3 / 18.6	13	17.1 / 17.6
	H	7	4.1 / 7.7	5	3.8 / 4.3	11	8.2 / 4.1	5	7.9 / 14.8	6	7.2 / 8.3	13	15.8 / 7.8

o: observed frequency.

f: fitted frequency (upper figure based on model containing $S \times I$, lower figure on independence model).

none of which is significant when referred to the tabulated χ^2-distribution. Atrium houses can in fact be fitted adequately by the independence model ($\chi^2 = 11.6$ with 12 d.f.), but show a suggested pattern of interaction similar to that in the other types of housing.

Apartments appear to require a model containing all two-factor interactions, although if only $S \times I$ as above is included the fit is considerably improved as compared with the main effects model. The remaining discrepancies can be judged in Table W.4 from the residuals, defined as

$$(\text{observed frequency} - \text{fitted frequency})/\sqrt{(\text{fitted frequency})}.$$

The largest residuals, 2.2 and -1.8, occur in apartments in compensating cells, i.e. the two cells corresponding to low and high contact within the same combination of low satisfaction/high influence; the fitted model gives a perfect fit within any particular satisfaction/influence combination. Thus, with the exception of this combination within apartments, each type of housing can be fitted by the same form of log linear model, i.e. one containing $S \times I$ but no other two-factor interaction.

Fitted frequencies under the model containing $S \times I$ are given in Table W.5. They agree closely with the observed frequencies apart from the instance already noted in apartments. Table W.5 gives also the fitted frequencies under

Table W.6. Ratios of frequencies under model containing $S \times I$ to frequencies under independence model

Satisfaction		Low	Medium	High
Housing	Influence			
Tower blocks				
	L	1.01	1.13	0.93
	M	1.20	1.04	0.88
	H	0.60	0.72	1.34
Apartments				
	L	1.46	1.03	0.57
	M	0.86	1.07	1.07
	H	0.58	0.86	1.47
Atrium houses				
	L	1.30	1.02	0.79
	M	0.80	1.08	1.07
	H	0.81	0.86	1.24
Terraced houses				
	L	1.26	0.88	0.64
	M	0.90	1.20	0.97
	H	0.53	0.88	2.02

the independence model. The meaning of the interaction is summarized more simply in Table W.6, in terms of ratios of the fitted frequencies under the model containing the interaction $S \times I$ to those under the independence model. Within each type of housing, increased satisfaction is associated with increased feeling of influence upon management. This holds even in atrium houses, for which the $S \times I$ interaction was earlier judged to be statistically nonsignificant. In particular, in the high-satisfaction/high-influence cell the numbers of respondents in tower blocks, apartments, atrium houses and terraced houses are 34, 47, 24 and 102 per cent, respectively, greater than would be the case under an assumed independence model.

Example X

Educational plans of Wisconsin schoolboys

Description of data. Sewell and Shah (1968) have investigated for some Wisconsin highschool 'senior' boys and girls the relationship between variables:

(i) socioeconomic status (high, upper middle, lower middle, low);
(ii) intelligence (high, upper middle, lower middle, low);
(iii) parental encouragement (low, high);
(iv) plans for attending college (yes, no).

The data for boys are given in Table X.1.

General considerations. A first crucial step is the choice of response variable or variables. We have studied the dependence of 'college plans' as a response

Table X.1. Socioeconomic status, intelligence, parental encouragement and college plans for Wisconsin schoolboys

IQ	College plans	Parental encouragement	SES			
			L	LM	UM	H
L	Yes	Low	4	2	8	4
		High	13	27	47	39
	No	Low	349	232	166	48
		High	64	84	91	57
LM	Yes	Low	9	7	6	5
		High	33	64	74	123
	No	Low	207	201	120	47
		High	72	95	110	90
UM	Yes	Low	12	12	17	9
		High	38	93	148	224
	No	Low	126	115	92	41
		High	54	92	100	65
H	Yes	Low	10	17	6	8
		High	49	119	198	414
	No	Low	67	79	42	17
		High	43	59	73	54

162

Example X 163

on the other variables as explanatory variables. Fienberg (1977, §7.3) used a more complicated approach regarding college plans as a final response variable partly to be explained by an intermediate response variable, parental encouragement. In effect, taking different response variables amounts to posing different questions and it is not to be supposed that there is just one allowable choice.

The first step is to calculate the proportions of boys responding 'Yes' to college plans for the $4 \times 4 \times 2$ classification of the explanatory variables. Inspection of these proportions is an essential step in descriptive analysis of the data. For more formal analysis it would be possible to apply analysis of variance directly to these proportions, ignoring changes in precision across the data and measuring effects directly in terms of differences of proportions. Partly because the proportions vary over a substantial range, the main analysis here has used a logistic model.

Table X.2. Proportions answering 'Yes' to college plans: observed and fitted proportions

IQ	Parental encouragement	SES			
		L	LM	UM	H
L	Low	0.01	0.01	0.05	0.08
		0.02	0.02	0.03	0.07
	High	0.17	0.24	0.34	0.41
		0.17	0.23	0.29	0.46
LM	Low	0.04	0.03	0.05	0.10
		0.03	0.04	0.06	0.12
	High	0.31	0.40	0.40	0.58
		0.27	0.35	0.42	0.61
UM	Low	0.09	0.09	0.16	0.18
		0.06	0.09	0.12	0.22
	High	0.41	0.50	0.60	0.78
		0.44	0.53	0.60	0.76
H	Low	0.13	0.18	0.12	0.32
		0.11	0.15	0.20	0.34
	High	0.53	0.67	0.73	0.88
		0.60	0.68	0.74	0.86

Upper figure in each pair is observed proportion, lower figure is fitted proportion in linear logistic model with only main effects.

The analysis. For the response variable college plans, Table X.2 gives the proportions answering 'Yes'. The proportions increase consistently with each of the explanatory variables.

The data are fitted adequately by a linear logistic model containing the three main effects, with no interactions; the difference in 2 log (maximized likelihood) between this and the full model giving perfect fit is 25.24 with 24 degrees of freedom. Each of the main effects is significant.

The estimated parameters and asymptotic standard errors for each of the main effects, calculated with GLIM, are given in Table X.3. Note that GLIM

Table X.3. Estimated parameters and standard errors

Parameter		Estimate	Standard error
SES	L	0	—
	LM	0.36	0.12
	UM	0.66	0.12
	H	1.41	0.12
IQ	L	0	—
	LM	0.59	0.12
	UM	1.33	0.12
	H	1.97	0.12
PE	L	0	—
	H	2.46	0.10

GLIM sets the first parameter of each group to zero.

sets the first parameter in any group to zero. Thus for socioeconomic status the estimated parameters are

$$L : 0, \quad LM : 0.36, \quad UM : 0.66, \quad H : 1.41,$$

i.e. the higher the socioeconomic status, the higher the proportion answering 'Yes' to college plans. The trend is even more marked across the levels of intelligence. For parental encouragement the estimated parameters are

$$L : 0, \quad H : 2.46.$$

Thus high parental encouragement increases the ratio 'Yes'/'No' by an estimated factor of $\exp(2.46) = 11.7$, with 95 per cent confidence limits 9.6 and 14.3.

The fitted proportions responding 'Yes' to college plans, based on the main-effects model, are shown in Table X.2 and are in close agreement with the observed proportions.

Related reference. Fienberg (1977, Chapters 6 and 7) gives a detailed discussion of the application of logistic models to cross-classified categorical data.

Summary of examples

'We summarize below the sets of data analysed in Part II and give also the main techniques used in the analysis.

Example A. Admissions to intensive care unit
Comparison with Poisson distribution via dispersion test. Intervals compared with exponential distribution. Simple time series analysis by forming subtotals.

Example B. Intervals between adjacent births
Simple analysis of three-way table.

Example C. Statistical aspects of literary style
Reduction of frequency distributions to summarizing statistics. Grouping of means into sets.

Example D. Temperature distribution in a chemical reactor
Regression. Prediction of proportion outside tolerance limit. Decomposition of variance.

Example E. A 'before and after' study of blood pressure
Detailed analysis of differences 'after' minus 'before'.

Example F. Comparison of industrial processes in the presence of trend
Formulation and fitting of linear model. Estimation of position of maximum.

Example G. Cost of construction of nuclear power plants
Multiple regression. Testing model adequacy. Model simplification.

Example H. Effect of process and purity index on fault occurrence
Binary data. Logistic models. Paired data. Comparison of maximized log likelihoods.

Example I. Growth of bones from chick embryos
Linear model for derived response. Recovery of between-block information.

165

Example J. Factorial experiment on cycles to failure of worsted yarn
Transformations. Analysis of variance. Partition of degrees of freedom.

Example K. Factorial experiment on diets for chickens
Analysis of variance. Main effects and interactions.

Example L. Binary preference data for detergent use
Factorial contrasts for unbalanced binary data. Analysis of variance.

Example M. Fertilizer experiment on growth of cauliflowers
Analysis of variance. Partition of degrees of freedom. Confounding.

Example N. Subjective preference data on soap pads
Ordinally scored data. Factorial experiment. Between- and within-block analyses.

Example O. Atomic weight of iodine
Unbalanced two-way analysis of variance.

Example P. Multifactor experiment on a nutritive medium
Unbalanced data. Choice of model out of many possible models. Multiple regression.

Example Q. Strength of cotton yarn
Components of variance. Allocation of observations.

Example R. Biochemical experiment on the blood of mice
Split-plot experiment. Factorial experiment.

Example S. Voltage regulator performance
Components of variance. Synthesis of variance. Prediction of proportion outside tolerance limit.

Example T. Intervals between failures of air-conditioning equipment in aircraft
Tests of exponential form. Fitting of sets of gamma distributions and associated maximized log likelihood tests. Adequacy of model.

Example U. Survival times of leukemia patients
Regression model for exponential distributions. Residuals.

Example V. A retrospective study with binary data
Estimation of effects from retrospective binary data. Examination of homogeneity.

Example W. Housing and associated factors
Comparisons involving three qualitative response variables. Log linear models.

Example X. Educational plans of Wisconsin schoolboys
Relation between binary response variable and several qualitative explanatory variables. Linear logistic model.

Further sets of data

Set 1

Gordon and Foss (1966) investigated the effect of rocking on the crying of very young babies; see also Cox (1970, p. 4). On each of 18 days, one baby in a hospital ward was selected at random and rocked. The other babies served as controls. At the end of a specified time the number of babies not crying were counted, with the results given in Table 1. Conditions, for example temperature, were appreciably different on different days.

Table 1. Numbers of control babies and experimental babies

Day	No. of control babies	No. not crying	No. of experimental babies	No. not crying
1	8	3	1	1
2	6	2	1	1
3	5	1	1	1
4	6	1	1	0
5	5	4	1	1
6	9	4	1	1
7	8	5	1	1
8	8	4	1	1
9	5	3	1	1
10	9	8	1	0
11	6	5	1	1
12	9	8	1	1
13	8	5	1	1
14	5	4	1	1
15	6	4	1	1
16	8	7	1	1
17	6	4	1	0
18	8	5	1	1

Set 2

Table 2 (Bissell, 1972) gives the numbers of faults in rolls of textile fabric. The distribution of number of faults is of interest, especially in its relation to that expected if faults occur at random at a fixed rate per metre.

Table 2. Numbers of faults in rolls of textile fabric

Roll No.	Roll length (metres)	No. of faults	Roll No.	Roll length (metres)	No. of faults
1	551	6	17	543	8
2	651	4	18	842	9
3	832	17	19	905	23
4	375	9	20	542	9
5	715	14	21	522	6
6	868	8	22	122	1
7	271	5	23	657	9
8	630	7	24	170	4
9	491	7	25	738	9
10	372	7	26	371	14
11	645	6	27	735	17
12	441	8	28	749	10
13	895	28	29	495	7
14	458	4	30	716	3
15	642	10	31	952	9
16	492	4	32	417	2

Set 3

Table 3 (J. S. Maritz, personal communication) gives data from an experiment on carcinogenesis in rats. Eighty rats were divided at random into 4 groups of 20 rats each, and treated as follows:

$$
\begin{array}{llll}
\text{Group I} & D, \text{ no } I, & \text{no } P; \\
\text{II} & D, & I, & \text{no } P; \\
\text{III} & D, \text{ no } I, & P; \\
\text{IV} & D, & I, & P;
\end{array}
$$

where D is thought to produce cancer, I is thought to act as an inhibitor and P is thought to accelerate the appearance of cancer. The data in Table 3 are survival times in days; after 192 days the experiment was ended, and a post mortem was conducted on every surviving rat to assess the presence or absence of cancer. In the table, 192^- means that the rat survived 192 days but was found to have cancer. The superscript $+$ means death from a cause unrelated to cancer; in particular, 192^+ means that on post mortem the rat did not have cancer.

Table 3. Survival times in days for four groups of rats

Group I; *D*		Group II; *DI*		Group III; *DP*		Group IV; *DIP*	
18+	106	2+	192+	37	51	18+	127
57	108	2+	192+	38	51	19+	134
63+	133	2+	192+	42	55	40+	148
67+	159	2+	192+	43+	57	56	186
69	166	5+	192+	43	59	64	192+
73	171	55+	192+	43	62	78	192+
80	188	78	192+	43	66	106	192+
87	192-	78	192-	43	69	106	192+
87+	192-	96	192-	48	86	106	192+
94	192-	152	192- .	49	177	127	192+

Set 4
The data in Table 4(a) relate to 47 states of the USA (Vandaele, 1978). The dependence of crime rate in 1960 (variable R) upon the other variables listed is of interest. The variables are defined in Table 4(b).

Table 4(a). Data on forty-seven states of the USA

R	Age	S	Ed	Ex_0	Ex_1	LF	M	N	NW	U_1	U_2	W	X
791	151	1	91	58	56	510	950	33	301	108	41	394	261
1635	143	0	113	103	95	583	1012	13	102	96	36	557	194
578	142	1	89	45	44	533	969	18	219	94	33	318	250
1969	136	0	121	149	141	577	994	157	80	102	39	673	167
1234	141	0	121	109	101	591	985	18	30	91	20	578	174
682	121	0	110	118	115	547	964	25	44	84	29	689	126
963	127	1	111	82	79	519	982	4	139	97	38	620	168
1555	131	1	109	115	109	542	969	50	179	79	35	472	206
856	157	1	90	65	62	553	955	39	286	81	28	421	239
705	140	0	118	71	68	632	1029	7	15	100	24	526	174
1674	124	0	105	121	116	580	966	101	106	77	35	657	170
849	134	0	108	75	71	595	972	47	59	83	31	580	172
511	128	0	113	67	60	624	972	28	10	77	25	507	206
664	135	0	117	62	61	595	986	22	46	77	27	529	190
798	152	1	87	57	53	530	986	30	72	92	43	405	264
946	142	1	88	81	77	497	956	33	321	116	47	427	247
539	143	0	110	66	63	537	977	10	6	114	35	487	166
929	135	1	104	123	115	537	978	31	170	89	34	631	165
750	130	0	116	128	128	536	934	51	24	78	34	627	135
1225	125	0	108	113	105	567	985	78	94	130	58	626	166
742	126	0	108	74	67	602	984	34	12	102	33	557	195
439	157	1	89	47	44	512	962	22	423	97	34	288	276
1216	132	0	96	87	83	564	953	43	92	83	32	513	227
968	131	0	116	78	73	574	1038	7	36	142	42	540	176
523	130	0	116	63	57	641	984	14	26	70	21	486	196
1993	131	0	121	160	143	631	1071	3	77	102	41	674	152
342	135	0	109	69	71	540	965	6	4	80	22	564	139
1216	152	0	112	82	76	571	1018	10	79	103	28	537	215
1043	119	0	107	166	157	521	938	168	89	92	36	637	154
696	166	1	89	58	54	521	973	46	254	72	26	396	237

Table (4a) contd.

R	Age	S	Ed	Ex0	Ex1	LF	M	N	NW	U1	U2	W	X
373	140	0	93	55	54	535	1045	6	20	135	40	453	200
754	125	0	109	90	81	586	964	97	82	105	43	617	163
1072	147	1	104	63	64	560	972	23	95	76	24	462	233
923	126	0	118	97	97	542	990	18	21	102	35	589	166
653	123	0	102	97	87	526	948	113	76	124	50	572	158
1272	150	0	100	109	98	531	964	9	24	87	38	559	153
831	177	1	87	58	56	638	974	24	349	76	28	382	254
566	133	0	104	51	47	599	1024	7	40	99	27	425	225
826	149	1	88	61	54	515	953	36	165	86	35	395	251
1151	145	1	104	82	74	560	981	96	126	88	31	488	228
880	148	0	122	72	66	601	998	9	19	84	20	590	144
542	141	0	109	56	54	523	968	4	2	107	37	489	170
823	162	1	99	75	70	522	996	40	208	73	27	496	224
1030	136	0	121	95	96	574	1012	29	36	111	37	622	162
455	139	1	88	46	41	480	968	19	49	135	53	457	249
508	126	0	104	106	97	599	989	40	24	78	25	593	171
849	130	0	121	90	91	623	1049	3	22	113	40	588	160

Table 4(b). Variables listed in Table 4(a)

The source is the *Uniform Crime Report* of the Federal Bureau of Investigation. All the data relate to calendar year 1960 except when explicitly stated otherwise.

R: Crime rate: the number of offences known to the police per 1 000 000 population.

Age: Age distribution: the number of males aged 14–24 per 1000 of total state population.

S: Dummy variable distinguishing place of occurrence of the crime (south = 1). The southern states are: Alabama, Arkansas, Delaware, Florida, Georgia, Kentucky, Louisiana, Maryland, Mississippi, North Carolina, Oklahoma, South Carolina, Tennessee, Texas, Virginia, and West Virginia.

Ed: Educational level: the mean number of years of schooling $\times 10$ of the population, 25 years old and over.

Ex_0, Ex_1:   Police expenditure: the per capita expenditure on police protection by state and local government in 1960 and 1959, respectively. Sources used are *Governmental Finances in 1960* and *Governmental Finances in 1959*, published by the US Bureau of the Census.

LF: Labour force participation rate per 1000 of civilian urban males in the age-group 14–24.

M: The number of males per 1000 females.

N: State population size in hundred thousands.

NW: Nonwhites: the number of nonwhites per 1000.

U_1:   Unemployment rate of urban males per 1000 in the age-group 14–24, as measured by census estimate.

U_2:   Unemployment rate of urban males per 1000 in the age-group 35–39.

W: Wealth as measured by the median value of transferable goods and assets or family income (unit 10 dollars).

X: Income inequality: the number of families per 1000 earning below one-half of the median income.

Set 5

Table 5 (Patterson and Silvey, 1980) gives the yield of six varieties of wheat in 1977 at ten testing centres in three regions of Scotland (N: North of Scotland, E: East of Scotland, W: West of Scotland). Not all six varieties were grown at each of the testing centres. The data were collected as part of a development and testing programme aimed at recommending new varieties. Patterson and Silvey estimated precision by detailed analysis of more extensive data: for the present purpose treat the standard error for one variety in one centre as known to be 0.19 (t grain/ha).

Table 5. Yield (t grain/ha) of winter wheat at ten centres

Variety	Centre									
	E1	E2	N3	N4	N5	N6	W7	E8	E9	N10
Huntsman	5.79	6.12	5.12	4.50	5.49	5.86	6.55	7.33	6.37	4.21
Atou	5.96	6.64	4.65	5.07	5.59	6.53	6.91	7.31	6.99	4.62
Armada	5.97	6.92	5.04	4.99	5.59	6.57	7.60	7.75	7.19	—
Mardler	6.56	7.55	5.13	4.60	5.83	6.14	7.91	8.93	8.33	—
Sentry	—	—	—	—	—	—	7.34	8.68	7.91	3.99
Stuart	—	—	—	—	—	—	7.17	8.72	8.04	4.70

Set 6

Healy *et al.* (1970) discuss an experiment in which each of 68 dental patients was given the anxiety-reducing drug diazepam immediately prior to a local anaesthetic and dental treatment. The patients formed two groups of equal size, a study group and a control group. The study group (*S*) were patients (with a high level of anxiety) who would not normally be willing to undergo a local anaesthetic and the control group (*C*) were patients (with a normal

Table 6. Average number of tasks completed in a set time by sixty-eight dental patients

Test	Group	Before drug	After drug		
			60 mins*	90 mins	1 week
I	C	64.8	50.2	64.6	66.7
	S	68.8	54.7	69.7	75.9
II	C	70.7	54.7	70.2	72.7
	S	65.7	45.7	66.2	68.7
III	C	78.5	58.5	81.8	83.3
	S	84.3	66.4	83.8	85.2

*Based on 20 patients.

anxiety level) who would normally be willing to do so. The drug enabled all the patients to accept a local anaesthetic. In order to examine the effect of the drug on motor co-ordination and dexterity, each patient performed three tests, I, II and III, (a) immediately before, (b) 60 minutes after, (c) 90 minutes after, and (d) one week after administration of the drug. Each test consisted of completing a simple task as many times as possible in a set time, the three tests involving different tasks. Each response in Table 6 is the average value, for all patients in the relevant group, of the number of tasks completed in the set time.

Set 7

Table 7, based on a survey in Fiji organized by World Fertility Survey (Little, 1978), gives the mean number of children born per woman, the women being classified by place, education and years since first marriage. Any systematic variation in number of children per woman is of interest.

Table 7. Mean number of children born to women in Fiji of Indian race, by marital duration, type of place and education. Observed mean values and sample sizes

Years since first marriage	Type of place							
	Urban				Rural			
	Education*				Education*			
	(1)	(2)	(3)	(4)	(1)	(2)	(3)	(4)
< 5	1.17	0.85	1.05	0.69	0.97	0.96	0.97	0.74
	12	27	39	51	62	102	107	47
5–9	4.54	2.65	2.68	2.29	2.44	2.71	2.47	2.24
	13	37	44	21	70	117	81	21
10–14	4.17	3.33	3.62	3.33	4.14	4.14	3.94	3.33
	18	43	29	15	88	132	50	9
15–19	4.70	5.36	4.60	3.80	5.06	5.59	4.50	2.00
	23	42	20	5	114	86	30	1
20–24	5.36	5.88	5.00	5.33	6.46	6.34	5.74	2.50
	22	25	13	3	117	68	23	2
25+	6.52	7.51	7.54	—	7.48	7.81	5.80	—
	46	45	13	0	195	59	10	0

*Categories of Education Level are:
(1) non, (2) lower primary, (3) upper primary, (4) secondary or higher.
Lower figures give the number of women involved.

Table 8. Mean annual malformation rates (per 1000 total singleton births), 1964–66

| Area | All CNS Malformations | | Anencephalus | | Spina bifida without anencephalus | | Water hardness (p.p.m.) |
	Non-manual	Manual	Non-manual	Manual	Non-manual	Manual	
Cardiff	4.62 (4 110)	8.21 (9 502)	1.22	3.26	2.19	3.47	110
Newport	5.25 (1 523)	5.18 (4 634)	0.66	0.65	4.60	3.24	100
Swansea	5.81 (2 408)	9.50 (5 579)	3.74	3.41	2.08	5.38	95
Glamorgan East Valleys	8.15 (3 189)	10.85 (13 362)	2.82	4.12	4.39	5.31	42
Glamorgan West Valleys	8.02 (1 995)	10.15 (8 279)	2.51	3.62	5.01	5.31	39
Rest of Glamorgan	5.14 (4 863)	8.26 (7 868)	2.26	3.18	2.47	3.56	161
Monmouthshire Valleys	7.56 (2 380)	8.56 (10 048)	2.52	3.58	3.36	3.68	83
Rest of Monmouthshire	5.58 (1 613)	7.50 (3 201)	1.86	2.50	3.72	4.06	122

Set 8

Lowe *et al.* (1971) investigated the association between hardness of the water supply and the incidence of births with malformation of the central nervous system (CNS), mainly anencephalus and spina bifida. Table 8 gives the mean annual malformation rates subdivided by social class (non-manual, manual) over the three years 1964–66 for eight areas in Wales; the total number of singleton births is shown in brackets. Births in which both anencephalus and spina bifida are present are counted under anencephalus, this being by far the more serious malformation.

Set 9

The data* in Table 9 relate to the comparison of 4 individual chemical processes carried out in a 4×4 Latin square arranged to eliminate variation between days and between times of the day. After a considerable time the experiment was repeated using a rerandomized form of the same design.

Table 9. Data on the yield from four chemical processes

Replicate 1	1	B:	3.44	D:	6.58	A:	8.21	C:	4.56
	2	C:	4.53	A:	8.83	B:	8.38	D:	8.11
Day	3	A:	7.38	C:	5.32	D:	8.52	B:	7.20
	4	D:	11.00	B:	11.78	C:	10.18	A:	13.61
Replicate 2	1	C:	3.60	A:	7.57	D:	10.22	B:	7.36
	2	B:	5.64	D:	7.46	C:	5.59	A:	10.63
Day	3	D:	9.30	B:	6.21	A:	11.55	C:	9.06
	4	A:	5.62	C:	5.72	B:	7.47	D:	11.13

*Fictitious data based on a real problem.

Set 10

Coleman (1964) describes a study in which 3398 US schoolboys were interviewed on two successive occasions. On each occasion it was recorded whether they were or were not members of the 'leading crowd' and their attitude to membership (+, −) was also noted, association between membership and attitude being of interest. The data are given in Table 10.

Table 10. Frequency of response according to membership and attitude to membership

		Second interview			
		Membership	+	+	−	−
		Attitude	+	−	+	−
First interview						
Membership	Attitude					
+	+		458	140	110	49
+	−		171	182	56	87
−	+		184	75	531	281
−	−		85	97	338	554

Set 11

The data in Table 11 (Ashford and Sowden, 1970) were collected as part of an investigation into the incidence of pneumoconiosis among coalminers. A sample of 18 282 coalminers who were known to be smokers but who showed no radiological abnormality were grouped by age and classified according to whether or not they showed signs of two symptoms, breathlessness and wheeze.

Table 11. Numbers of subjects responding to breathlessness and wheeze according to age group

Breathlessness		Yes		No		Total
Wheeze		Yes	No	Yes	No	
	20–24	9	7	95	1 841	1 952
	25–29	23	9	105	1 654	1 791
	30–34	54	19	177	1 863	2 113
	35–39	121	48	257	2 357	2 783
Age group (years)	40–44	169	54	273	1 778	2 274
	45–49	269	88	324	1 712	2 393
	50–54	404	117	245	1 324	2 090
	55–59	406	152	225	967	1 750
	60–64	372	106	132	526	1 136
Total		1 827	600	1 833	14 022	18 282

Set 12

In an attempt to assess the effect of attitude towards capital punishment on the voting behaviour of jurors in criminal cases, 464 jurors who had voted in split ballots were interviewed (Zeisel, 1968). The three key questions asked were:

(i) Do you have any conscientious scruples against the death penalty?
(ii) How did you vote on the first ballot after the jury started to deliberate?
(iii) How did the entire jury vote on this first ballot?

The data in Table 12 give, for each possible split of the jury, the numbers of jurors with and without scruples against the death penalty who voted (a) Guilty, (b) Undecided, and (c) Not Guilty, on the first ballot. Interest lies in whether jurors without scruples against the death penalty are more likely to vote Guilty on the first ballot than jurors who have such scruples.

Table 12. Votes of 464 jurors

Jury split		No. of jurors voting		
Guilty/Not Guilty	Scruples	Not Guilty	Undecided	Guilty
1/11	without	2	0	1
	with	5	0	2
2/10	without	24	2	1
	with	10	3	5
3/9	without	14	1	6
	with	14	0	2
4/8	without	6	1	6
	with	16	2	9
5/7	without	1	0	3
	with	4	0	1
6/6	without	9	8	10
	with	6	2	5
7/5	without	7	1	11
	with	4	1	3
8/4	without	9	0	26
	with	7	1	18
9/3	without	3	1	24
	with	7	1	22
10/2	without	3	2	36
	with	4	1	29
11/1	without	0	0	29
	with	3	1	29

Set 13
The data in Table 13 are taken from a report on a cohort study into radiation upon mortality among survivors of the Hiroshima atom bomb (Otake, 1979). Of particular interest is the incidence of death from leukemia.

Table 13. Number of individuals alive in 1950 and deaths during the period 1950–59

Age at 1950		Radiation dose in rads						
		Total	0	1–9	10–49	50–99	100–199	200+
5–14	Leukemia	14	3	1	0	1	3	6
	All other cancers	2	1	0	0	0	0	1
	Other causes	141	48	41	28	11	6	7
	Alive 1950	15 286	6 675	4 084	2 998	700	423	406
15–24	Leukemia	15	0	2	3	1	3	6
	All other cancers	13	6	4	2	0	0	1
	Other causes	392	195	101	46	10	22	18
	Alive 1950	17 109	7 099	4 716	2 668	835	898	893
25–34	Leukemia	10	2	2	0	0	1	5
	All other cancers	27	9	9	4	2	1	2
	Other causes	290	122	80	52	15	10	11
	Alive 1950	10 424	4 425	2 646	1 828	573	459	493
35–44	Leukemia	8	0	0	1	1	1	5
	All other cancers	114	55	30	17	2	2	8
	Other causes	418	179	99	76	20	22	22
	Alive 1950	11 571	5 122	2 806	2 205	594	430	414
45–54	Leukemia	20	9	3	2	0	1	5
	All other cancers	328	127	81	73	21	11	15
	Other causes	990	452	229	197	48	35	29
	Alive 1950	12 472	5 499	3 004	2 392	664	496	417
55–64	Leukemia	10	2	0	2	2	1	3
	All other cancers	371	187	80	57	22	17	8
	Other causes	1 403	635	362	256	53	53	44
	Alive 1950	8 012	3 578	2 011	1 494	434	283	212
65+	Leukemia	3	1	1	0	0	0	1
	All other cancers	256	119	59	48	13	10	7
	Other causes	2 264	1 039	604	418	99	63	41
	Alive 1950	4 862	2 245	1 235	935	232	123	92

Set 14

Ten nominally similar perfumes for use in a disinfectant were assessed as follows. Ten small measured samples of the disinfectant, differing only in respect of the perfume, were placed in ten identical containers. A panel of 30 judges were each asked to smell the samples and to group them according to whether or not any difference in smell could be detected; the number of groups could be any number from one (no detectable difference) to ten (all different). Each panel member was also asked to classify each group of samples according to acceptability of perfume into 'like', 'indifferent' or 'dislike'.

The data are given in Table 14. Samples judged to smell alike are bracketed together; thus, the first judge allocated the 10 samples into 4 groups, the second into 8 groups, etc. The identification of any groups in which the perfumes appear to be indistinguishable is of importance, and also the acceptability of the perfume in these groups.

Set 15

In an experiment carried out in four sections on the assessment of meat tenderness (Bouton *et al.*, 1975), subjective and objective measurements were obtained on samples of beef. The sections of the experiment were done under

Table 14. Allocation by thirty judges of ten perfumes into groups

Judge	Acceptability Like	Indifferent	Dislike
1	(2, 5, 8) (1, 4, 6, 9, 10)		(3) (7)
2	(1) (6) (7)	(3)	(2, 4, 5) (8) (9) (10)
3	(1, 2, 4, 5, 6, 9, 10)	(8)	(3) (7)
4	(3, 7, 10)		(5, 6, 8) (1, 2, 4, 9)
5	(5) (6) (9)	(1) (2) (4) (7) (8)	(3) (10)
6	(1, 4, 7, 8) (5) (6) (9) (10)		(2) (3)
7	(7)	(5) (4, 8, 9, 10)	(3) (1, 2, 6)
8		(1) (4) (5) (6) (10)	(2, 3, 9) (7, 8)
9	(1, 4, 5, 7, 8, 10) (2, 6, 9)		(3)
10	(1, 2, 5, 6) (7, 8, 9, 10)		(3, 4)
11	(1, 4, 8, 9, 10) (2, 5, 6, 7)		(3)
12	(8) (4, 6) (9, 10)	(5) (7) (1, 2)	(3)
13	(2, 7, 9) (5, 8, 10)	(1) (4, 6)	(3)
14		(2) (5) (7) (9)	(1, 4, 8, 10) (3, 6)
15	(1, 4, 9) (2, 5, 6, 7, 8, 10)		(3)
16	(1, 4, 10)	(2) (5) (6) (7) (8) (9)	(3)
17	(5, 6, 8)		(1, 2, 7, 9, 10) (3, 4)
18	(1, 4, 7, 9)	(2, 5, 6, 8, 10)	(3)
19	(2, 3, 8)	(5, 6, 10)	(1, 4, 7, 9)
20	(1) (2) (4) (5) (6) (7) (8) (9) (10)		(3)
21	(2, 4, 6, 9) (8, 10)	(5, 7) (1)	(3)
22	(2, 6, 8, 10)	(1, 4, 5, 9)	(3) (7)
23	(4, 9) (6, 10)		(3, 7) (1, 2, 5, 8)
24	(1, 9, 10)	(2, 5) (4, 6, 8)	(3, 7)
25		(7) (1, 2, 5) (4, 6, 8, 9, 10)	(3)
26	(5, 10)	(1, 2, 6, 9)	(3, 7) (4, 8)
27	(1, 4, 7, 9)	(2, 5, 6, 8, 10)	(3)
28	(1, 2, 6)	(5, 9, 10)	(3, 4, 7, 8)
29	(2) (3, 5, 6, 10)		(1, 4, 7, 8, 9)
30	(1, 4)	(3, 5)	(2, 7) (6, 8) (9, 10)

different cooking conditions and with varied cuts of beef with the objective of producing appreciable differences in tenderness. A twelve-member taste panel subjectively assessed the tenderness and juiciness on a scale 0–15 (0 = extremely tender or juicy, 15 = very tough or very dry). Objective measurements were made of (i) force required to compress a sample to 80 per cent of its thickness, (ii) force required to pull a sample apart, (iii) shear force and (iv) percentage weight lost in cooking. The data are summarized in Table 15. The objective measurements given are averages of 6–10 observations, the standard errors being based on the variation within groups. The aim is to examine the extent to which the variation in tenderness and juiciness can be accounted for by the variations in the objective measurements.

Table 15. Mean values of objective measurements obtained for each treatment and mean panel scores for samples bitten across the fibres (A) and between the fibres (B)

Section	Sample	Objective				Subjective			
		Compression*	Adhesion*	Shear*	Cooking loss %	Tenderness (A)	(B)	Juiciness (A)	(B)
1	1	2.29	0.74	7.42	39.0	12.0	9.4	7.3	7.3
	2	1.44	0.38	6.74	41.7	9.6	7.4	8.9	8.5
	3	1.14	0.24	6.13	42.7	8.0	5.9	9.2	9.3
	4	0.81	0.14	5.29	43.1	8.2	4.1	9.4	9.8
	Standard error	0.09	0.05	0.74	0.4	0.7	0.7	0.3	0.3
2	5	2.69	1.32	6.65	39.8	11.8	9.8	8.2	8.9
	6	2.49	0.96	7.31	36.3	9.5	9.0	6.9	7.2
	7	1.99	0.49	8.22	36.9	9.4	7.8	7.7	7.3
	8	1.26	0.17	4.36	32.1	4.7	3.1	5.5	5.9
	Standard error	0.09	0.09	0.46	0.5	0.4	0.4	0.3	0.3
3	9	1.92	0.63	6.36	20.3	7.3	6.5	4.2	3.4
	10	2.59	0.97	7.13	27.3	9.5	9.1	6.4	4.6
	11	2.41	1.01	7.35	36.6	10.8	8.8	9.6	8.7
	12	2.02	0.79	7.83	41.0	9.6	7.3	11.2	10.3
	Standard error	0.14	0.07	0.52	0.5	0.3	0.3	0.3	0.3
4	13	1.93	0.63	13.28	35.8	12.1	9.6	8.4	7.5
	14	1.82	0.68	7.79	36.0	8.0	7.5	8.9	8.5
	15	1.78	0.64	8.22	13.7	7.8	5.5	2.4	2.2
	16	1.57	0.64	5.45	17.1	4.5	3.7	3.3	2.7
	Standard error	0.05	0.04	0.35	0.8	0.2	0.2	0.4	0.3

*All values in kg

References

Anderson, S., Auquier, A., Hauck, W.W., Oakes, D., Vandaele, W. and Weisberg, H.I. (1980). *Statistical Methods for Comparative Studies*. New York: Wiley.

Armitage, P. (1971). *Statistical Methods in Medical Research*. Oxford: Blackwell.

Ashford, J.R. and Sowden, R.R. (1970). Multi-variate probit analysis. *Biometrics*, **26**, 535–46.

Baxter, G.P. and Landstredt, O.W. (1940). A revision of the atomic weight of iodine. *J. Amer. Chem. Soc.*, **62**, 1829–34.

Biggers, J.D. and Heyner, S. (1961). Studies on the amino acid requirements of cartilaginous long bone rudiments *in vitro*. *J. Exp. Zool.*, **147**, 95–112.

Bishop, Y.M.M., Fienberg, S.E. and Holland, P.W. (1975). *Discrete Multivariate Analysis: Theory and Practice*. Cambridge, Mass: MIT Press.

Bissell, A.F. (1972). A negative binomial model with varying element sizes. *Biometrika*, **59**, 435–41.

Bouton, P.E., Ford, A.L., Harris, P.V. and Ratcliff, D. (1975). Objective–subjective assessment of meat tenderness. *J. Texture Stud.*, **6**, 315–28.

Box, G.E.P. and Cox, D.R. (1964). An analysis of transformations (with discussion). *J.R. Statist. Soc.*, B, **26**, 211–52.

Brownlee, K.A. (1965). *Statistical Theory and Methodology in Science and Engineering*. 2nd edn. New York: Wiley.

Cochran, W.G. and Cox, G.M. (1957). *Experimental Designs*, 2nd edn. New York: Wiley.

Coleman, J.S. (1964). *Introduction to Mathematical Sociology*. New York: The Free Press of Glencoe.

Cox, D.R. (1958). *Planning of Experiments*. New York: Wiley.

Cox, D.R. (1970). *Analysis of Binary Data*. London: Chapman and Hall.

Cox, D.R. and Lewis, P.A.W. (1966). *The Statistical Analysis of Series of Events*. London: Chapman and Hall.

Davies, O.L. (1963). *Design and Analysis of Industrial Experiments*, 2nd edn. London: Longman.

Davies, O.L. and Goldsmith, P.L. (1972). *Statistical Methods in Research and Production*, 4th edn. London: Longman.

Desmond, D.J. (1954). Quality control on the setting of voltage regulators. *Applied Statist.*, **3**, 65–73.

Draper, N.R. and Smith, H. (1981). *Applied Regression Analysis*, 2nd edn. New York: Wiley.

Fedorov, V.D., Maximov, V.N. and Bogorov, V.G. (1968). Experimental development of nutritive media for micro-organisms. *Biometrika*, **55**, 43–51.

Feigl, P. and Zelen, M. (1965). Estimation of exponential survival probabilities with concomitant information. *Biometrics*, **21**, 826–38.

Fienberg, S.E. (1977). *The Analysis of Cross-Classified Categorical Data*. Cambridge, Mass: MIT Press.

Gordon, T. and Foss, B.M. (1966). The role of stimulation in the delay of onset of crying in the new-born infant. *J. Exp. Psychol.*, **16**, 79–81.

Greenberg, R.A. and White, C. (1963). The sequence of sexes in human families. Paper presented to the 5th International Biometric Conference, Cambridge.

Healy, T.E.J., Lautch, H., Hall, N., Tomlin, P.J. and Vickers, M.D. (1970). Inter-disciplinary study of diazepam sedation for outpatient dentistry. *Brit. Med. J.*, **3**, 13–17.

John, J.A. and Quenouille, M.H. (1977). *Experiments: Design and Analysis*, 2nd edn. London: Griffin.

Johnson, N.L. (1967). Analysis of a factorial experiment. (Partially confounded 2^3). *Technometrics*, **9**, 167–70.

Little, R.J.A. (1978). Generalized linear models for cross-classified data from the W.F.S. *Technical Bulletin Series* No. 5/TECH. 834, International Statistical Institute.

Lowe, C.R., Roberts, C.J. and Lloyd, S. (1971). Malformations of central nervous system and softness of local water supplies. *Brit. Med. J.*, **2**, 357–61.

MacGregor, G.A., Markandu, N.D., Roulston, J.E. and Jones, J.C., (1979). Essential hypertension: effect of an oral inhibitor of angiotension-converting enzyme. *Brit. Med. J.*, **2**, 1106–9.

Madsen, M. (1976). Statistical analysis of multiple contingency tables. Two examples. *Scand. J. Statist.*, **3**, 97–106.

Morton, A.Q. (1965). The authorship of Greek prose (with discussion). *J.R. Statist. Soc.*, A, **128**, 169–233.

Mooz, W.E. (1978). Cost analysis of light water reactor power plants. *Report R-2304-DOE*. Rand Corp., Santa Monica, Calif.

Otake, M. (1979). Comparison of time risks based on a multinomial logistic response model in longitudinal studies. *Technical Report No. 5*, RERF, Hiroshima, Japan.

Patterson, H.D. and Silvey, V. (1980). Statutory and recommended list trials of crop varieties in the U.K. (with discussion). *J. R. Statist. Soc.*, A, **143**, 219–52.

Pearson, E.S. and Hartley, H.O. (1966). *Biometrika Tables for Statisticians*, Vol. 1, 3rd ed. Cambridge University Press.

Pearson, E.S. and Hartley, H.O. (1972). *Biometrika Tables for Statisticians*, Vol. 2. Cambridge University Press.

Proschan, F. (1963). Theoretical explanation of observed decreasing failure rate. *Technometrics*, **5**, 375–83.

Ries, P.N. and Smith, H. (1963). The use of chi-square for preference testing in multidimensional problems. *Chem. Eng. Progress*, **59**, 39–43.

Seber, G.A.F. (1977). *Linear Regression Analysis*. New York: Wiley.

Sewell, W.H. and Shah, V.P. (1968). Social class, parental encouragement and educational aspirations. *Amer. J. Sociol.*, **73**, 559–72.

Snedecor, G.W. and Cochran, W.G. (1967). *Statistical Methods*, 6th edn. Ames, Iowa: Iowa State Univ. Press.

Vandaele, W. (1978). Participation in illegitimate activities: Ehrlich revisited, in *Deterrence and Incapacitation*, Blumstein, A., Cohen, J. and Nagin, D. (eds), pp. 270–335. Washington, D.C.: National Academy of Sciences.

Wetherill, G.B. (1967). *Elementary Statistical Methods*. London: Chapman and Hall.

Woolf, B. (1955). On estimating the relation between blood group and disease. *Ann. Hum. Genetics*, **19**, 251–3.

Zeisel, H. (1968). *Some data on juror attitudes towards capital punishment*. University of Chicago Law School.

Author Index

Anderson, S., 76, 181
Armitage, P., 57, 94, 125, 138, 154, 181
Ashford, J.R., 176, 181
Auquier, A., 181

Barella, A., 98
Barr, A., 53
Baxter, G.P., 121, 181
Biggers, J.D., 95, 181
Bishop, Y.M.M., 111, 181
Bissell, A.F., 169, 181
Bogorov, V.G., 126, 181
Bouton, P.E., 178, 181
Box, G.E.P., 102, 181
Brownlee, K.A., 121, 181

Carpenter, K., 103
Cochran, W.G., 57, 94, 95, 102, 125, 138, 142, 154, 181, 182
Coleman, J.S., 175, 181
Cox, D.R., 57, 94, 102, 111, 135, 138, 147, 154, 168, 181
Cox, G.M., 95, 181

Davies, O.L., 57, 102, 115, 181
Desmond, D.J., 139, 181
Draper, N.R., 81, 90, 181
Duckworth, J., 103

Fedorov, V.D., 126, 181
Feigl, P., 148, 182
Fienberg, S.E., 111, 163, 164, 181, 182
Ford, A.L., 181
Foss, B.M., 168, 182

Goldsmith, P.L., 57, 181
Gordon, T., 168, 182
Greenberg, R.A., 58, 182
Gower, J.C., 112

Hall, N., 182
Harris, P.V., 181

Hartley, H.O., 61, 80, 133, 142, 182
Hauck, W.W., 181
Healy, T.E.J., 172, 182
Heyner, S., 95, 181
Holland, P.W., 111, 181

John, J.A., 103, 106, 182
Johnson, N.L., 116, 182
Jones, J.C., 72, 182

Landstredt, O.W., 121, 181
Lautch, H., 182
Lewis, P.A.W., 57, 147, 181
Little, R.J.A., 173, 182
Lloyd, S., 182
Lowe, C.R., 175, 182

MacGregor, G.A., 72, 73, 182
Madsen, M., 155, 182
Maritz, J.S., 169
Markandu, N.D., 72, 182
Maximov, V.N., 126, 181
Mooz, W.E., 81, 182
Morton, A.Q., 63, 182

Oakes, D., 181
Otake, M., 177, 182

Patterson, H.D., 172, 182
Pearson, E.S., 61, 80, 133, 142, 182
Proschan, F., 143, 182

Quenouille, M.H., 103, 106, 182

Ratcliff, D., 181
Ries, P.N., 107, 111, 182
Roberts, C.J., 182
Roulston, J.E., 72, 182

Seber, G.A.F., 81, 182
Sewell, W.H., 162, 182
Shah, V.P., 162, 182
Silvey, V., 172, 182

185

Subject Index

187